W9-CHA-013

Cook College
16 Cooper St
Westmont, N J
08108

Woody Plants in Winter

A Manual of Common Trees and Shrubs
in Winter in the Northeastern United States
and Southeastern Canada

by

EARL L. CORE
Department of Biology
West Virginia University

and

NELLE P. AMMONS
Department of Biology
West Virginia University

Pacific Grove, California

© 1958
by
The Boxwood Press

Reprinted with corrections, 1973

No part of this book may be reproduced
in any form without written permission
of the Copyright owner.

Distributed by

THE BOXWOOD PRESS
183 Ocean View Blvd.
Pacific Grove, CA 93950

Paper edition, $2.95
Cloth edition, $4.25

SBN: 910286-02-7

Cover photograph
by
John A. Gibson
Professor of Chemistry
West Virginia University

Printed in U.S.A.

Preface

THE IDENTIFICATION OF WOODY PLANTS IN WINTER has attracted the attention of students of botany for many years. While at first glance trees and shrubs may seem remarkably alike in their dormant condition it is soon discovered that many of their winter characters provide just as satisfactory criteria for recognition as do the characters exhibited during the growing season.

This manual is the outgrowth of many years of teaching by the authors during which time several editions of a winter manual were prepared and used in classes. The present treatment embodies improvements which have resulted from actual use in taxonomy classes over the years.

The area covered by this manual is, in general, the northeastern United States and southeastern Canada. The limits might be indicated as the southern boundaries of Virginia and Kentucky, the western boundaries of Missouri and Iowa, and the 49th parallel of latitude through Quebec and Ontario to the northwestern corner of Minnesota. In general, ranges are stated from east to west across the north, then to the south. However, in the case of some distinctly southern plants having the principal portion of their range south of this area the opposite practice has been followed. No attempt has been made to include all woody plants of the area, but most of the important species of the various regions are treated. Introduced plants are included if they are likely to occur spontaneously.

The Latin names, in general, conform to those used in Gray's Manual of Botany, 8th Edition. Other names that have been in frequent use by various authors are included in parenthesis.

Most of the pen-and-ink drawings have been made by Nelle P. Ammons, some of them having appeared in previous works by the present authors. The sketches in the opening chapter were made by Dr. William A. Lunk of the University of Michigan.

PREFACE

The writers acknowledge indebtedness to numerous individuals and institutions for assistance in the preparation of this work. Particular gratitude is expressed to the staff of the Arnold Arboretum of Harvard University for their kindness in furnishing material for study. Appreciation is also extended to Dr. P. D. Strausbaugh, Professor Emeritus of Botany, West Virginia University for his helpful suggestions. The cover photograph,of trees in winter, was made by Dr. John A. Gibson, Professor of Chemistry, West Virginia University.

THE AUTHORS

June, 1958

Contents

WOODY PLANTS IN WINTER

THE IDENTIFICATION of woody plants in winter has a well-developed technique of its own which, while quite different from the techniques used for the determination of the same plants during the other seasons of the year, may be fully as reliable. Most woody plants have an individuality of their own which is just as evident in winter, at least upon careful examination, as it is in summer. The morphological features used are, of course, not usually those of flowers or fruits, or even of leaves, but the size and form of the plants and especially the structures of the younger branches, or twigs.

Size and Form of Woody Plants. The sizes of woody plants are, of course, quite variable, but an effort is made, in the descriptions of the following pages, to give the general range of the height (and for trees, especially, the trunk diameter) of mature plants. It is realized that trees, when young, are no taller than shrubs, and it may be difficult in winter to tell if a given plant has reached maturity or not, but the figures would be meaningless otherwise.

Trees and shrubs grown in the open tend to have a characteristic form well illustrated by such species as white pine, Lombardy poplar, weeping willow, and American elm. In general, however, under forest conditions the crown is restricted in its development by the competition of surrounding trees and the form is not so distinctive.

Bark. Bark, or periderm, is that portion of a woody stem which lies outside of the cambium layer, and is protective in nature. The very youngest stems are covered by epidermis, but this is usually quickly succeeded by periderm, having its origin in superficial layers of the cortex. Compared with older bark, this is quite smooth, broken only by lenticels. At this stage, color is an important feature for diagnostic purposes. On older branches or trunks cork cambium (Fig. 1) originates deeper in the stem and when these deeper layers are formed the tissues on the outside die and the bark tends to become scaly or furrowed (Fig. 4), often according a very distinctive pattern. In sour gum, flowering dogwood, and persimmon, for example, the bark is divided into small

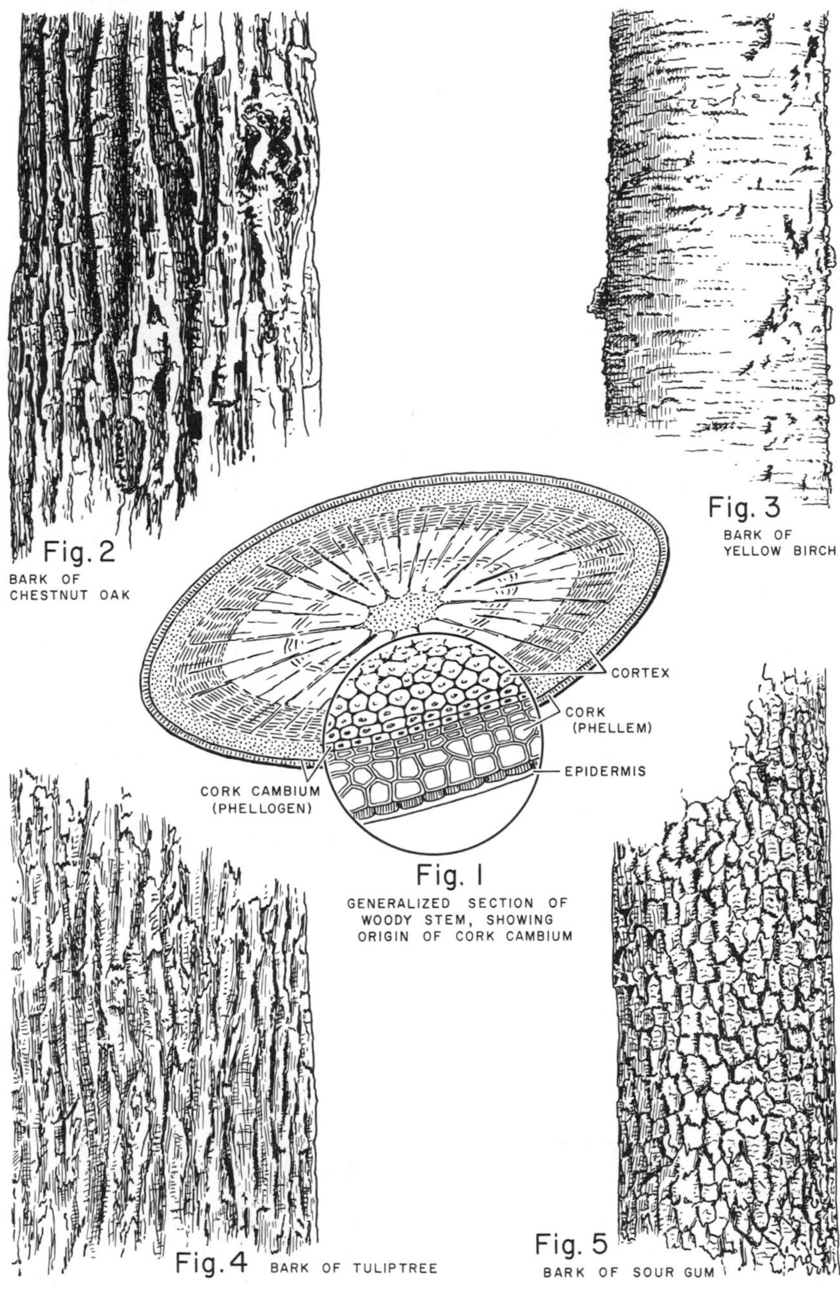

Fig. 2
BARK OF CHESTNUT OAK

Fig. 3
BARK OF YELLOW BIRCH

Fig. 1
GENERALIZED SECTION OF WOODY STEM, SHOWING ORIGIN OF CORK CAMBIUM

Fig. 4 BARK OF TULIPTREE

Fig. 5
BARK OF SOUR GUM

squarish plates and has the general appearance of alligator leather (Fig. 5). In cherries and birches the bark peels horizontally into thin sheets (Fig. 3). Broad exfoliating plates characterize certain hickories. Deeply furrowed bark is found on many trees, as chestnut oak (Fig. 2). Lumbermen often rely almost wholly upon bark characteristics in the identification of timber species, but the features are often difficult to describe so they can be recognized. Experience is most important in this type of identification.

TWIGS

The terminal portion of a branch of a woody plant may be referred to as a branchlet or twig (Figs. 6, 7). In this work these terms are used to designate, specifically, the growth of the current year and, to a lesser extent, that of the last preceding year. Twigs bear prominent distinguishing features such as buds, leaf-scars, stipule-scars, and pith, while their color, taste, and odor may also be distinctive. Various other features, such as corky ridges, thorns, and pubescence characterize certain species.

Color and hairiness of twigs are important winter characters, although they might more properly be regarded as physiological effects rather than physical features.

During the winter the color tends to become darker on the side most exposed to the sun (the upper, or southern sides). Twigs which are green in summer may become reddish in winter through the formation of anthocyanins, favored by cold weather. Thereafter the color deepens each year, for two or three or even four years. Hairs also change somewhat in color, as well as in abundance as the winter season progresses. Twigs become less and less hairy in the second and later years, under ordinary conditions. Of course some are glabrous from the beginning.

Buds. A bud, literally, is a growing point, the early undeveloped stage of a leafy shoot or a flower. In winter, of course, the growing points are dormant, and are usually covered, for protection, by bud scales, which are really modified leaves.

Buds are of two types, the terminal buds at the tips of the stem and its branches, and the lateral buds, along the sides of the branchlets (Fig. 6). In some species a true terminal bud is not formed and growth continues throughout the season until stopped by unfavorable weather. When this has happened, the young tender tip of the shoot dies back and is finally self-pruned at the highest mature lateral bud formed. This bud then appears to be terminal and is referred to as a pseudoterminal bud. Usually it continues the growth of the shoot in the next season, in much the same manner as an actual terminal bud would have done. A pseudoter-

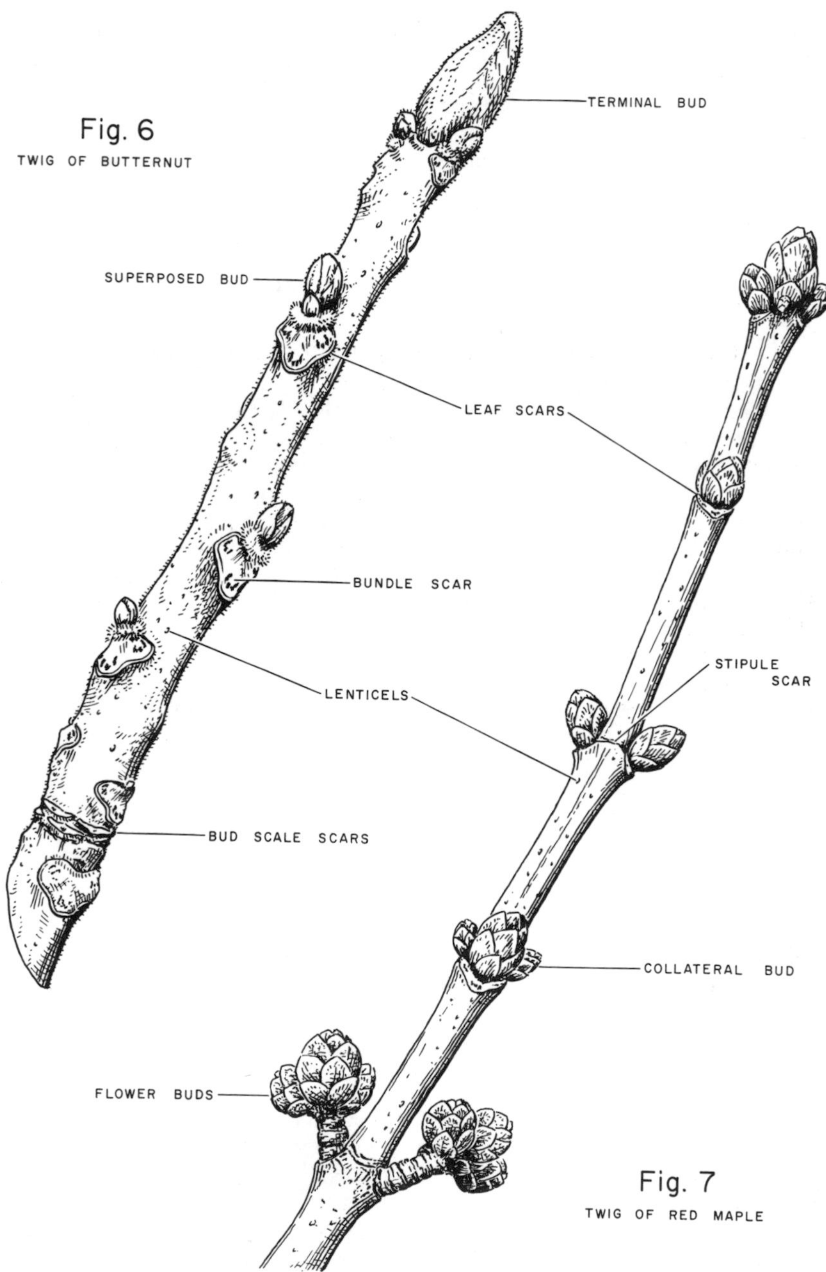
Fig. 6
TWIG OF BUTTERNUT
TERMINAL BUD
SUPERPOSED BUD
LEAF SCARS
BUNDLE SCAR
STIPULE SCAR
LENTICELS
BUD SCALE SCARS
COLLATERAL BUD
FLOWER BUDS
Fig. 7
TWIG OF RED MAPLE

minal bud may be distinguished from a true terminal by the fact that it usually has a leaf-scar immediately below it, and a small twig-scar behind it. The twig-scar, as might be expected, shows the concentric zones of bark, wood, and pith, characteristic of most woody stems. Rarely a withered twig tip persists and no clear-cut twig scar is formed.

Lateral buds are called axillary buds if they arise in the axils of leaves, as they usually do. The axil is the distal angle formed by the petiole of the leaf with the shoot. Often more than one bud appears at a node, in which case the one directly above the leaf scar is considered the true axillary bud and the others are designated as accessory buds. Accessory buds produced to the right or left of the axillary bud are said to be collateral (Fig. 7), while those produced just above the axillary bud are said to be superposed (Fig. 6). Often the accessory buds may be flower buds, whereas the axillary bud might be a leaf bud. In other cases both flowers and leaves are borne together in mixed buds. Buds differ greatly in their size and shape, as well as in the number, arrangement, color, size, shape, and surface nature of the bud scales; all these are valuable taxonomic features. When the scales of a bud fall as spring growth begins, they leave on the twig a ring of bud scale scars (Fig. 6). A series of such scars indicates several years' growth.

Bud scales, as noted above, are actually modified leaves (or, rarely, stipules), and serve to protect the enclosed embryonic structures. In some buds the scales may be rather numerous, overlapping each other like the shingles of a house; such an arrangement is said to be imbricate. In other cases the scales (generally 2 in number) do not overlap but fit together edge to edge; these are valvate scales. Of course, in both imbricate and valvate buds, the exposed scales may not represent the total number; others may be concealed, becoming exposed only in spring, when the bud enlarges as growth begins. In a few plants, as willows, the buds are covered by a single scale. Not all buds are scaly; some lack scales and are referred to as naked buds. These, however, have the actual growing point protected by less strongly modified, rudimentary leaves which often show veins and are generally scurfy or pubescent.

Leaf scars. Most woody plants of the northeastern United States are deciduous, i.e., the leaves fall as the growing season comes to a close. The fall of the leaf is associated with the development of a corky abscission layer and after the leaf has fallen there remains at the point of its attachment a portion of this layer, known as the leaf scar, sealing off the living tissues beneath (Fig. 8). Since the petioles vary greatly in appearance in cross section, the leaf scars are also quite variable, and are of further taxo-

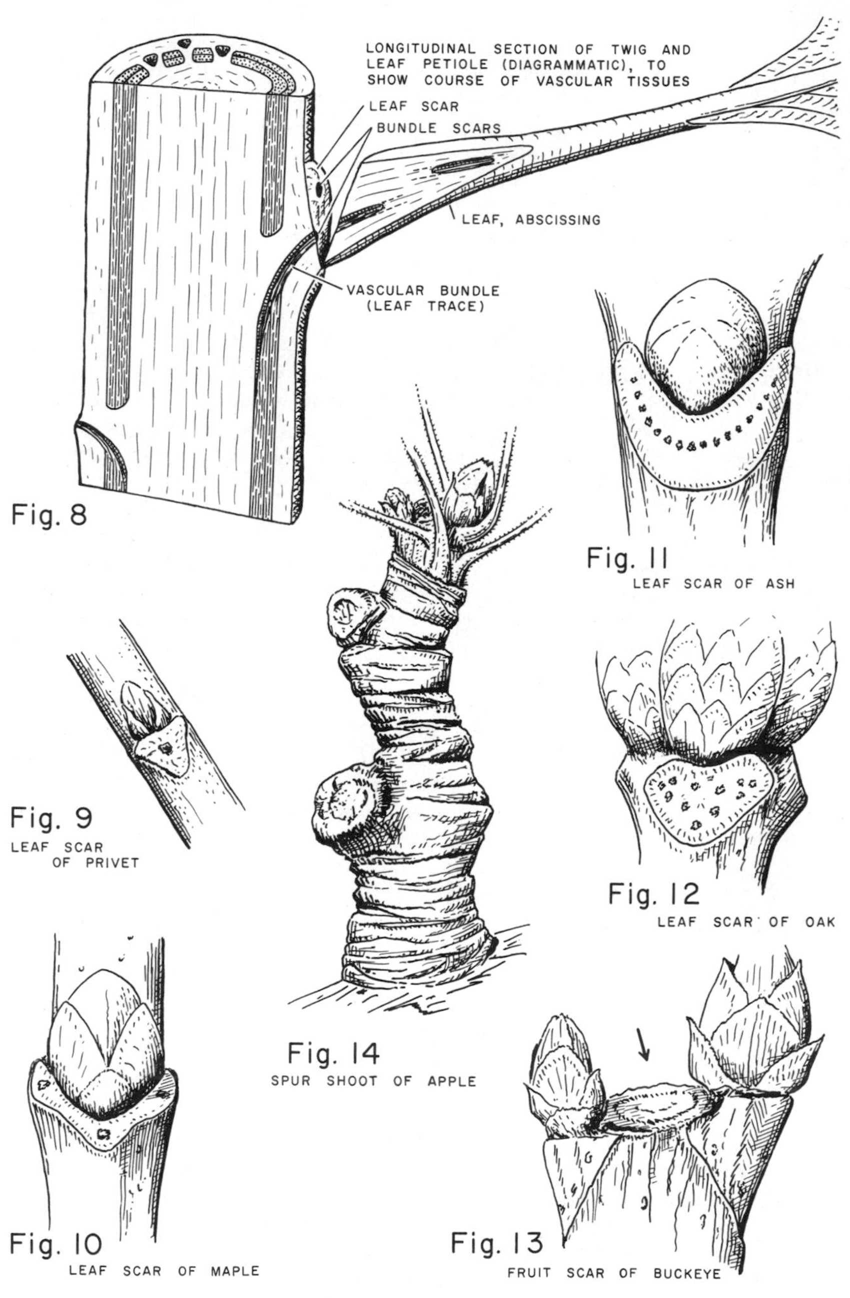

Fig. 8

Fig. 11
LEAF SCAR OF ASH

Fig. 9
LEAF SCAR OF PRIVET

Fig. 12
LEAF SCAR OF OAK

Fig. 14
SPUR SHOOT OF APPLE

Fig. 10
LEAF SCAR OF MAPLE

Fig. 13
FRUIT SCAR OF BUCKEYE

nomic value because of the variation in number and arrangement of the bundle scars (or traces), which indicate the broken ends of the vascular bundles passing from the stem into leaves. Common numbers of traces are one or three (Figs. 9, 10, 11, 12), although the usual number for a given species may be increased through compounding or branching of the bundles before they reach the abscission layer. Tiny stipule scars may also be present, one on each side of the leaf scars, marking the points where the stipules were attached (Fig. 15), or the modified stipules themselves may be present in the form of paired prickles (black locust, Fig. 17) bud scales (magnolia), or tendrils (greenbrier). Of course, stipules or stipule scars are not found on all twigs, since some species of plants are exstipulate (without stipules). The scars, when present, are generally slit-like in shape and inconspicuous. In a few species, as sycamore and tuliptree, (Fig. 16), they encircle the twig from one edge of the leaf scar to the other.

Branch and Fruit Scars. As noted above, some species do not form a true terminal bud, but the withered tip of the shoot may slough off, leaving a branch scar. In a few species, e.g. buffalo-nut, short lateral branches bearing several leaves may drop at the close of the season; in such species branch scars may sometimes be more numerous than leaf scars.

Fruit scars are similar in appearance to branch scars, but are often found in a terminal position (Fig. 13). They represent the point of abscission of the shoot bearing the inflorescence and the fruits. In species which normally have a true terminal bud, as buckeye, the presence of fruit scars may be mystifying until their real nature is learned.

Pith. The central portion of a twig is composed of a cylinder of parenchyma cells called pith. It is usually a different color from the xylem (wood) surrounding it and is readily recognizable in transverse or longitudinal sections of twigs. In most species the pith is circular in cross section but it may be star-shaped (oaks), 5-sided (cottonwoods), or more or less triangular (alders). In color it is usually white but may be various shades of pink, yellow, brown, or green.

Usually the pith is continuous and homogeneous in composition (Fig. 18). A modification of this type is diaphragmed, having plates of heavier-walled horizontally-elongated cells at more or less regularly spaced intervals (Fig. 19). Examples are tuliptree and black gum. In a few species the pith disappears between the diaphragms, resulting in small empty spaces; such pith is called chambered (Fig. 20). Pith may also be spongy, i.e., filled with small irregular cavities, or partially or entirely excavated, i.e., "dug out" or lacking.

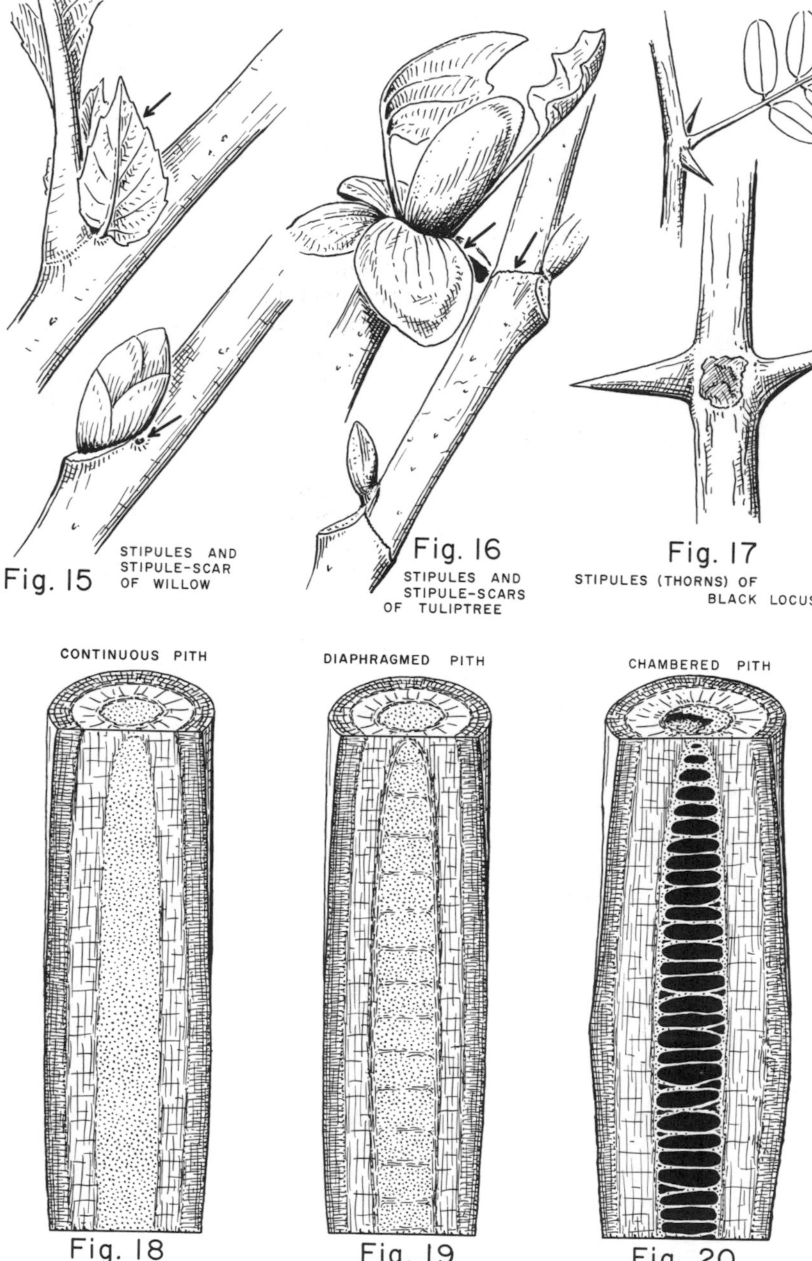

Fig. 15 STIPULES AND STIPULE-SCAR OF WILLOW

Fig. 16 STIPULES AND STIPULE-SCARS OF TULIPTREE

Fig. 17 STIPULES (THORNS) OF BLACK LOCUST

Fig. 18

Fig. 19

Fig. 20

Studies of the pith may be facilitated by application of a small drop of phloroglucin, followed by a drop of hydrochloric acid. This results in the wood turning a bright red in color, presenting the outline of the pith in sharp focus.

Lenticels. Lenticels are small, often wart-like prominences scattered over the surface of twigs; they serve to admit air to the living tissues beneath (Figs. 6, 7). They may be circular in shape or quite irregular. In cherries and birches they are elongated horizontally. In some cases they are relatively conspicuous, as in elder, in other cases quite inconspicuous. In general they are of little value in identification of twigs.

Spur Shoots. In some species (e.g., larch, birch) certain twigs grow very slowly and appear as dwarf branches, even though they may bear a normal number of leaves. A spur shoot is short, usually stocky, and with crowded leaf scars. Often the flower buds may be produced on spurs, as in apple and pear (Fig. 14).

Prickles and thorns. Small spines or prickles occur on the twigs of numerous species of woody plants and they are often diagnostic in character. In some cases they represent modified leaves, as in barberry. In other cases they are modified stipules, as in black locust (Fig. 17), while in still other cases they are cortical emergences (outgrowths of the cortex), scattered over the surface of the twig, as in rose and gooseberry.

Thorns, on the other hand, are modified sharp-pointed twigs, and have the vascular bundles characteristic of other twigs. In some cases careful examination will reveal the presence of tiny buds and leaf scars. They may be branched, as in honey-locust, or unbranched, as in hawthorn. In some instances (e.g., crabapple), a twig may be sharp-pointed, without actually appearing to be a thorn and might represent a structure from the evolutionary standpoint on the way to becoming a thorn.

FRUITS AND FLOWERS

Fruits. A true fruit is a ripened ovary, bearing one or more ripened ovules, the seeds. The wall of the fruit, developed from the wall of the ovary, is called the pericarp. This pericarp may be fleshy, relatively soft and juicy, or dry, relatively hard and tough; of course all sorts of gradations exist between these two types.

Fruits are quite often present on woody plants in winter. These are most likely to be dry fruits, since fleshy fruits are more perishable and are likely to have fallen and rotted, or to have been eaten by animals, particularly birds. Those fleshy fruits which do persist until winter are likely to have withered, as the coralberry, or to be fruits with scanty pulp, as the hackberry.

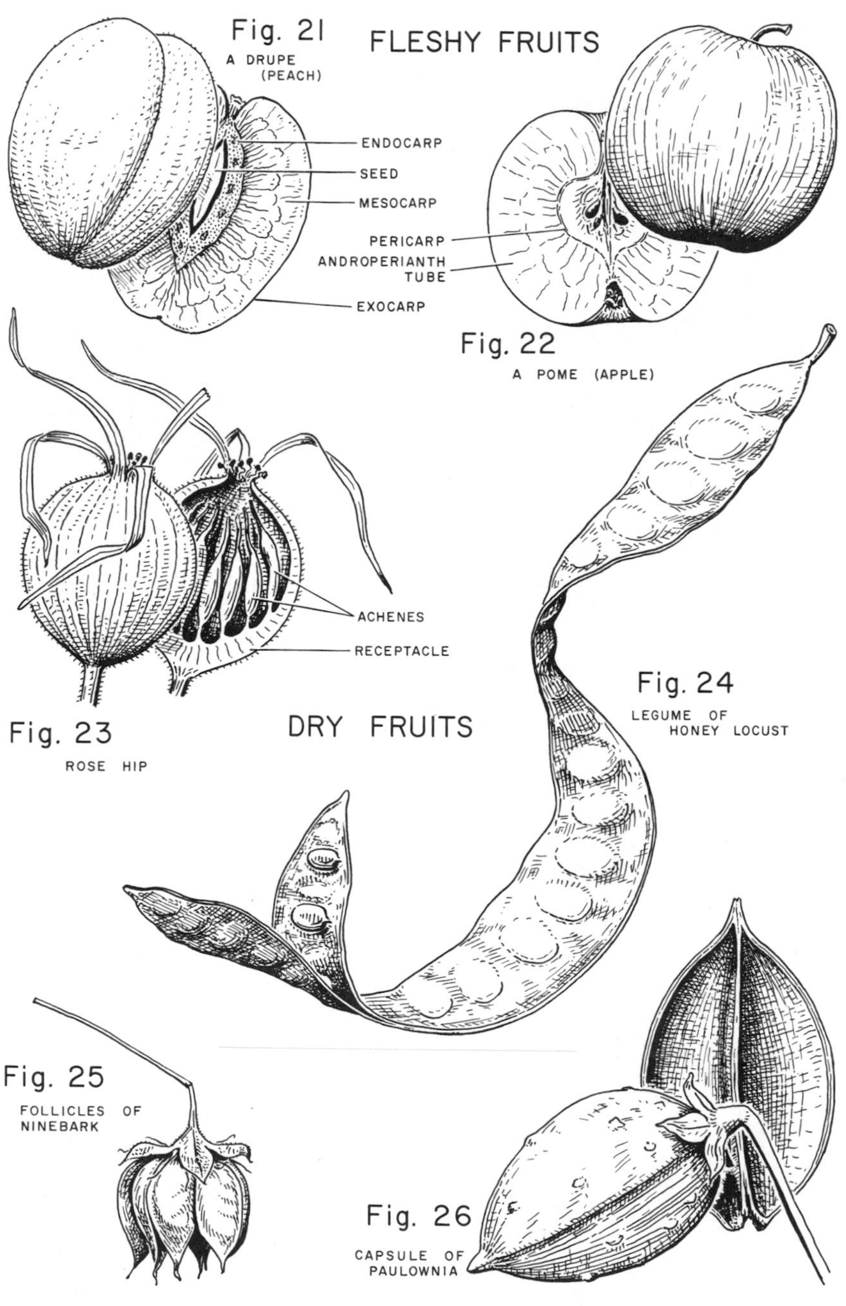
Fig. 21
A DRUPE
(PEACH)
FLESHY FRUITS
ENDOCARP
SEED
MESOCARP
PERICARP
ANDROPERIANTH
TUBE
EXOCARP
Fig. 22
A POME (APPLE)
ACHENES
RECEPTACLE
Fig. 24
LEGUME OF
HONEY LOCUST
Fig. 23
ROSE HIP
DRY FRUITS
Fig. 25
FOLLICLES OF
NINEBARK
Fig. 26
CAPSULE OF
PAULOWNIA

Of fleshy fruits present in winter, one of the most common is the drupe (Fig. 21). This has the pericarp consisting of three distinct layers, the exocarp, mesocarp, and endocarp, the endocarp taking the form of a hard stone surrounding the seed; fruits of this type are often called "stone fruits". Examples are black haw and hackberry. Drupes are usually considered as one-seeded, but in some instances, as in holly, several stones (endocarps), each with its own seed, may be surrounded by the same mesocarp. The pome (Fig. 22) is developed from a several-carpelled, several-seeded inferior ovary, the fleshy portion being a combination of pericarp and androperianth tube (a tube composed of the coalesced lower portions of the sepals, petals, and stamens); examples are hawthorn, chokeberry, and mountain-ash. The rose hip (Fig. 23) is an aggregate fruit, having many separate carpels of a single flower within the fleshy, hollowed-out receptacle.

Dry fruits include the legume (e.g., redbud), a fruit with one locule but splitting along two sutures (Fig. 24); the follicle (e.g., ninebark), like a legume but splitting along one suture only; (Fig. 25); and the capsule (e.g., catalpa), a dry fruit of two or more carpels, splitting into each locule (Fig. 26). All these may be classed as dehiscent (opening to permit the escape of numerous seeds). Other dry fruits are indehiscent containing but one seed and therefore not needing to open, since the entire fruit functions as a single seed. Among dry indehiscent fruits are the achene, a small fruit with the pericarp closely investing the seed (e.g., sycamore, with the many achenes compounded to form a globose head); the samara (Figs. 27, 28, 29), like an achene but provided with a wing which favors wind dispersal (e.g., elm, hoptree); and the nut (Figs. 30, 31), like an achene but larger, with a hard leathery or bony pericarp (e.g., chestnut, beech). A special type of a nut is the acorn of oaks, where the nut proper is borne in a cup representing a modified involucre (Fig. 32). In walnut and hickory the nut has some of the properties of a drupe, with an exocarp and a mesocarp (the husk) separating from the hard bony endocarp (the shell); such a fruit has been called a tryma (Fig. 33).

FLOWERS. Very few woody plants produce their flowers during the winter months. Witch-hazel, the only fall-blooming shrub in our region, sometimes has its yellow flowers persisting until winter (Fig. 38). Most winter-blooming plants, however, are in reality pre-vernal (early spring) in nature, coming into bloom in late February or March, and usually persisting into spring (or in the event of unusually sustained cold weather, not beginning to bloom until spring). Examples are willows, elms, and maples (Figs. 34, 35, 36, 37, 39).

GYMNOSPERM "FRUITS"—The seeds of conifers or cone-bearing plants (pines, spruces, firs, etc.) are borne exposed on the

DRY FRUITS

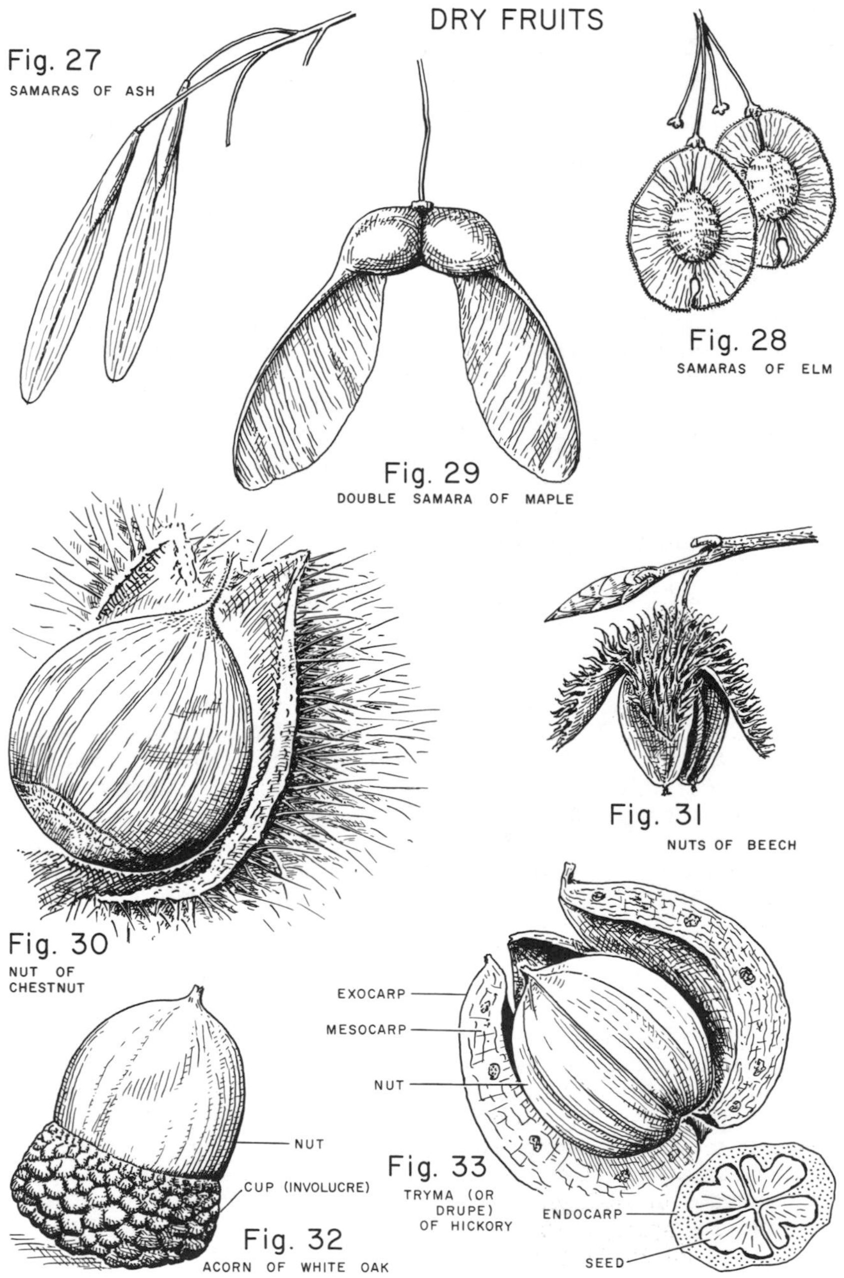

Fig. 27
SAMARAS OF ASH

Fig. 28
SAMARAS OF ELM

Fig. 29
DOUBLE SAMARA OF MAPLE

Fig. 30
NUT OF CHESTNUT

Fig. 31
NUTS OF BEECH

Fig. 32
ACORN OF WHITE OAK

Fig. 33
TRYMA (OR DRUPE) OF HICKORY

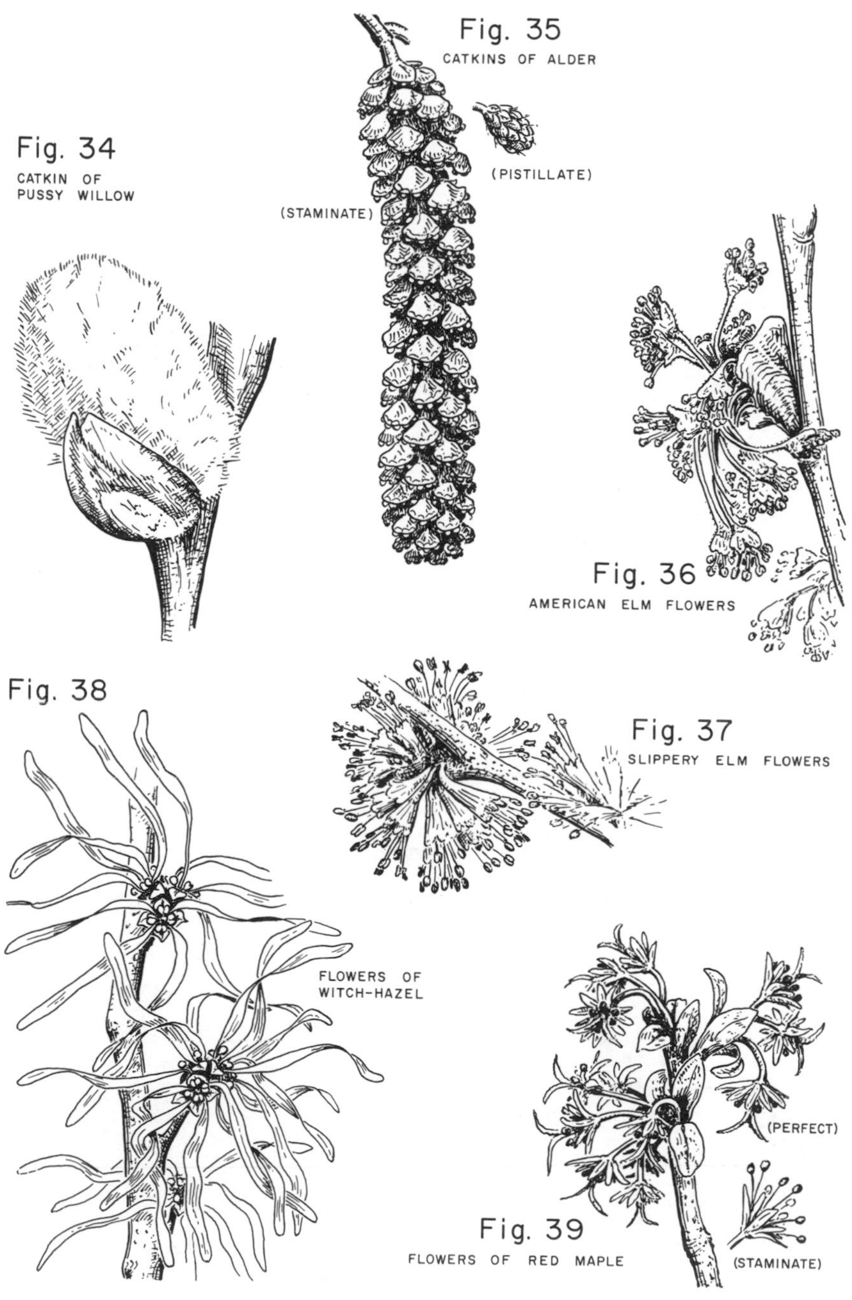

Fig. 34
CATKIN OF PUSSY WILLOW

Fig. 35
CATKINS OF ALDER

Fig. 36
AMERICAN ELM FLOWERS

Fig. 37
SLIPPERY ELM FLOWERS

Fig. 38

Fig. 39
FLOWERS OF RED MAPLE

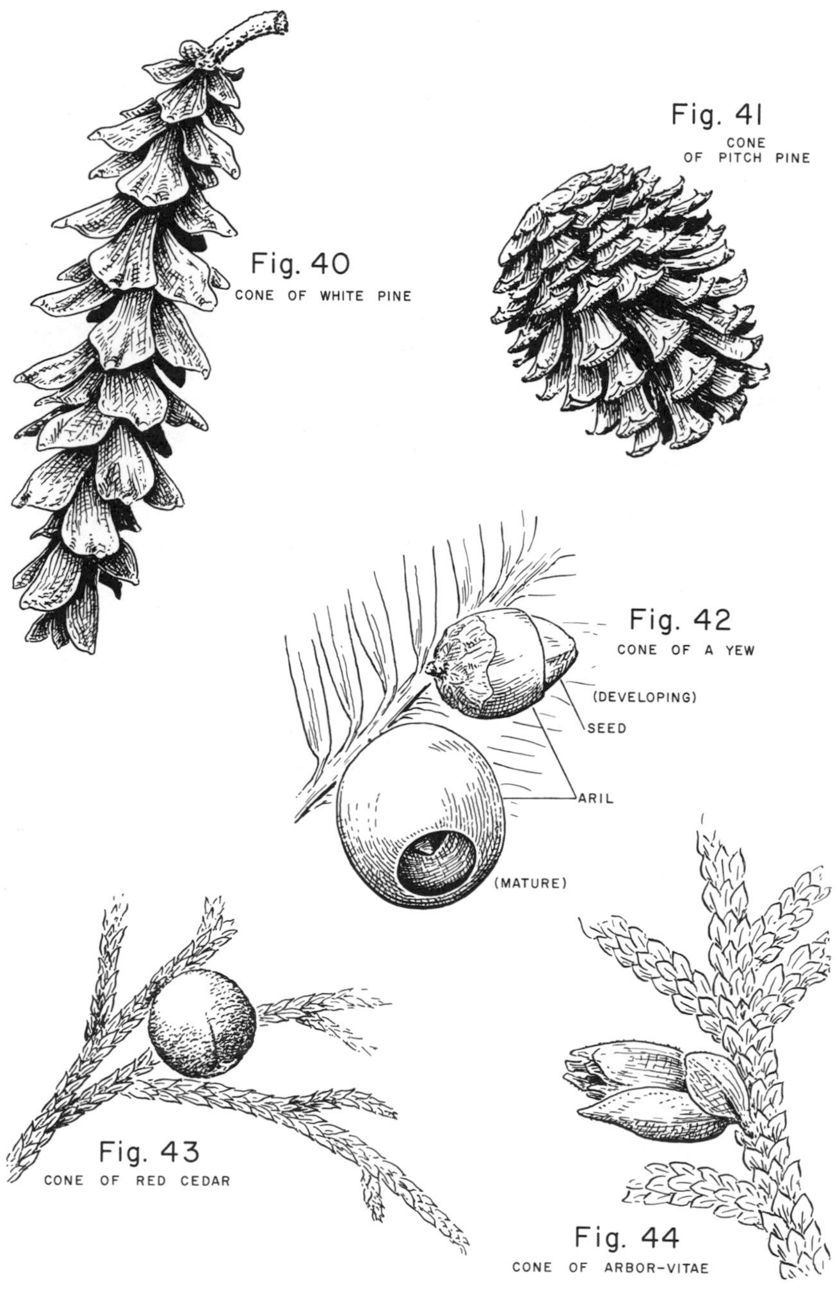

Fig. 40 CONE OF WHITE PINE

Fig. 41 CONE OF PITCH PINE

Fig. 42 CONE OF A YEW

Fig. 43 CONE OF RED CEDAR

Fig. 44 CONE OF ARBOR-VITAE

surface of the cone scales (hence Gymnosperm, "naked seed") and not enclosed in an ovary, as in the case of Angiosperms ("covered seeds"). If a fruit be defined as a ripened ovary, there is, therefore, no such thing as a true fruit in a gymnosperm. Nevertheless, the cones are generally regarded as fruits, and will be so treated here.

The cones of many conifers are available in winter, either attached to the branches or scattered on the ground under the tree. They are, therefore, quite useful in winter identification. Like the fruits of Angiosperms, coniferous fruits may be dry or fleshy. In yew, the solitary seed is partially or wholly surrounded by a fleshy aril (Fig. 42). In pines and most other conifers of our region, there are numerous woody, leathery, or fleshy scales, each with one or more seeds, generally arranged about a central axis to form a cone (Fig. 40, 41). The cones generally remain closed while the seeds are developing, but open at maturity to permit the seeds to escape. The exposed portion of the cone scale in an unopened cone is known as the apophysis. In some species the apophysis is smooth, in others wrinkled, ridged, or grooved. When the cone opens, the apophysis can generally be identified because it is lighter in color than the portions of the cone which had not been exposed.

In several species the apophysis terminates in a small scar called the umbo. When the umbo is located at the tip of the scale, it is said to be terminal; if it is on the back, it is dorsal in its location. Umbos may terminate in prickles or spines, which are diagnostic and vary with the species, being straight or curved, weak or strong.

In the Pinaceae the scales are spirally inserted, but in the Cupressaceae (Thuja, Juniperus) they are opposite and decussate (Fig. 44). In most conifers the scales are woody or leathery, but in red cedar they are somewhat fleshy and the cone is berry-like in appearance (Fig. 43).

KEY TO THE GENERA OF WOODY PLANTS

1. Foliage evergreen. 2
1. Foliage deciduous. 31

FOLIAGE EVERGREEN

2. Small yellowish-green parasites on trees... Phoradendron, p. 89
2. Not as above. 3

3. Leaves 8 mm. or less in width. 4
3. Leaves 12 mm. or more in width. 15

4. Leaves in fascicles of 2, 3, or 5... Pinus, p. 35
4. Leaves not in fascicles. 5

5. Leaves alternate, in spirals, linear. 6
5. Leaves opposite, awl-shaped or scale-like. 13

6. Leaves 3 - or 4 - angled in cross section... Picea, p. 33
6. Leaves flattened. 7

7. Stems trailing or creeping. 8
7. Stems erect. 9

8. Stems chaffy; leaves green on both sides... Gaultheria, p. 173
8. Stems not chaffy; leaves whitened beneath... Vaccinium, p. 175

9. Low sprawling shrubs. 10
9. Trees. 12

10. Leaves white-hairy... Hudsonia, p. 154
10. Leaves glabrous. 11

11. Leaves alternate; shrubs to 1 m. high... Taxus, p. 31
11. Leaves opposite; usually less than 25 cm. high... Pachistima, p. 137

12. Leaves sessile... Abies, p. 31
12. Leaves jointed to short persistent sterigmata... Tsuga, p. 31

13. Sprays not flattened. 14
13. Sprays flattened... Thuja, p. 41

14. Leaves opposite, scale-like... Chamaecyparis, p. 43
14. Leaves dimorphic, ternate on young growth, opposite on older branches... Juniperus, p. 43

15. Leaves opposite... Lonicera, p. 187
15. Leaves alternate. 16

16. Climbing plants. 17
16. Not climbing. 18

17. Climbing by tendrils... Smilax, p. 45
17. Climbing by aerial roots... Hedera, p. 158

18. Stems creeping or low (under 20 cm. high). 23
18. Stems much taller. 19

19. Fruit an acorn... Quercus, p. 74
19. Fruit not an acorn. 20

20. Stipule-scars encircling the twig... Magnolia, p. 94
20. Not as above. 21

21. Leaves very rusty-woolly beneath... Ledum, p. 163
21. Not as above. 22

22. Leaves scurfy-pubescent beneath... Chamaedaphne, p. 171
22. Not as above. 24

23. Stems red-bristly; fruit dry... Epigaea, p. 171
23. Stems glabrate; fruit fleshy, red... Gaultheria, p. 173

24. Leaves whitened beneath... Andromeda, p. 169
24. Leaves not conspicuously whitened beneath. 25

25. Leaves entire. 26
25. Leaves not entire. 29

26. Fruits persistent, in several-flowered racemes... Pieris, p. 169
26. Not as above. 27

27. Leaves 3 cm. or less long... Arctostaphylos, p. 173
27. Leaves more than 3 cm. long. 28

28. Buds moderate or large, scales several... Rhododendron, p. 163
28. Buds minute, scales 2 or none... Kalmia, p. 167

29. Leaves armed with spiny teeth... Ilex, p. 134
29. Leaves minutely serrulate. 30

30. Leaves less than 3 cm. long... Gaylussacia, p. 175
30. Leaves up to 15 cm. long... Leucothoe, p. 170

FOLIAGE DECIDUOUS

31. Leaf-and stipule-scars lacking, the buds subtended by minute scales... Taxodium, p. 41
31. Not as above. 32

32. Leaf-scars opposite or whorled. 33
32. Leaf-scars alternate. 89

33. Leaf-scars whorled. 34
33. Leaf-scars opposite. 38

Leaf-Scars Whorled

34. Twigs coarse...Catalpa, p. 186
34. Twigs slender. 35

35. Buds small, often sunk in bark...Cephalanthus, p. 186
35. Buds medium-sized, very evident. 36

36. Twigs with two or four hairy lines below each node...Diervilla, p. 187
36. Not as above. 37

37. Soft-wooded shrubs...Hydrangea, p. 99
37. Trees (Abnormal specimens)..Acer, p. 139

Leaf-Scars Opposite

38. Climbing or scrambling. 39
38. Not climbing. 42

39. Climbing by tendrils or aerial roots. 40
39. Twining or scrambling, without tendrils or aerial roots...Lonicera, p. 187

40. Climbing by tendrils. 41
40. Climbing by aerial roots...Campsis, p. 185

41. Section of stem showing wood in shape of a cross...Bignonia, p. 185
41. Wood not cross-shaped in section...Clematis, p. 91

42. Some twigs ending in spines...Rhamnus, p. 143
42. Twigs not ending in spines. 43

43. Bundle-traces distinct, three or more in a line. 44
43. Bundle-traces one or many; if many, scattered or in a ring or nearly confluent in a line. 76

44. Leaf-scars large and broad. 45
44. Leaf-scars small or narrow. 48

45. Trees. 46
45. Soft-wooded shrubs. 47

46. Twigs with milky juice...Acer, p. 139
46. Twigs without milky juice...Aesculus, p. 143

47. Lenticels very conspicuous...Sambucus, p. 195
47. Lenticels not conspicuous...Hydrangea, p. 99

48. Leaf-scars linear, straight or curved. 49
48. Leaf-scars not as above. 61

49. Leaf-scars horseshoe-shaped...Calycanthus, p. 97
49. Leaf-scars straight or curved, but not horseshoe-shaped. 50

50. Bud-scale one. 51
50. Bud-scales two or more or none. 52

51. Buds gummy within...Viburnum, p. 191
51. Buds not gummy within...Salix, p. 46

52. Bud-scales none (the young foliage naked, rusty-brown)...Viburnum, p. 191
52. Buds with two or more scales. 53

53. Bud-scales two, valvate. 54
53. Bud-scales more than two. 58

54. Youngest growth gray, or grayish-brown, not hairy. 55
54. Youngest growth often highly colored (red, green, purple, etc.). 57

55. Leaf-buds long, slender, tapering, lead-colored...Viburnum, p. 191
55. Leaf-buds not long and slender. 56

56. Buds dark red or rusty-brown, scurfy...Viburnum, p. 191
56. Buds gray or brown, not scurfy...Cornus, p. 158

57. Leaf-scars V-shaped, not raised, nearly meeting...Acer, p. 139
57. Leaf-scars U-shaped, somewhat raised, connected on each side by a narrow line...Cornus, p. 158

58. Buds sessile. 59
58. Buds stalked. 60

59. Branches short, numerous, rigidly spreading...Viburnum, p. 191
59. Not as above...Acer, p. 139

60. Leaf-scars meeting in a point...Acer, p. 139
60. Leaf-scars not meeting...Viburnum, p. 191

61. Leaf-scars very small and inconspicuous; pith often more or less excavated...Lonicera, p. 187
61. Leaf-scars larger, pith continuous. 62

62. Buds covered. 63
62. Buds not covered. 64

63. Buds covered by the base of the petiole... Cornus, p. 158
63. Buds covered by a broad membrane... Philadelphus, p. 98

64. Leaf-scars ciliate at top. 65
64. Leaf-scars not ciliate. 66

65. Buds sessile... Acer, p. 139
65. Buds stalked... Viburnum, p. 191

66. Twigs with two or four raised hairy lines... Diervilla, p. 187
66. Twigs without raised lines. 67

67. Bud-scales none... Viburnum, p. 191
67. Buds with scales. 68

68. Exposed scales two or four. 69
68. Exposed scales more than four. 72

69. Buds stalked... Acer, p.139
69. Buds not stalked. 70

70. Leaf-scars connected by lines... Lonicera, p. 187
70. Leaf-scars not connected by lines. 71

71. Stipule-scars present, almost connecting the leaf scars... Staphylea, p. 139
71. Stipule-scars none... Acer, p. 139

72. Buds superposed. 73
72. Buds not superposed. 74

73. Leaf-scars raised... Lonicera, p. 187
73. Leaf-scars low... Diervilla, p. 187

74. Soft-wooded shrubs... Hydrangea, p. 99
74. Trees or shrubs, not soft-wooded. 75

75. Leaf-scars meeting in a point... Acer, p. 139
75. Leaf-scars not meeting... Viburnum, p. 191

76. Bundle-traces many. 77
76. Bundle-trace one. 79

77. Sap milky... Broussonetia, p. 87
77. Sap not milky. 78

78. Pith continuous... Fraxinus, p. 179
78. Pith chambered... Paulownia, p. 185

79. Scurfy with peltate scales... Shepherdia, p. 155
79. Not as above. 80

80. Pith chambered or excavated. 81
80. Pith continuous. 83

81. Buds fusiform, multiple... Forsythia, p. 181
81. Not as above. 82

82. Twigs green, somewhat 4-sided... Euonymus, p. 137
82. Twigs terete, hairy, brown or gray... Symphoricarpos, p. 189

83. Soft-wooded shrubs with flaking bark... Hypericum, p. 153
83. Not as above. 84

84. Leaf-scars 2 mm. or more broad. 85
84. Leaf-scars less than 2 mm. broad. 88

85. Buds superposed. 86
85. Buds not superposed... Syringa, p. 181

86. Leaf-scars connected by lines... Cephalanthus, p. 186
86. Leaf-scars not connected by lines. 87

87. Shrubs with conspicuous lenticels... Chionanthus, p. 183
87. Trees with moderate lenticels... Fraxinus, p. 179

88. Leaf-scars shriveled... Symphoricarpos, p. 189
88. Leaf-scars distinctly outlined... Ligustrum, p. 183

Leaf-Scars Alternate

89. Climbing or scrambling. 90
89. Not climbing. 103

90. Climbing by tendrils or aerial roots. 91
90. Twining or scrambling, without aerial roots or tendrils. 95

91. With tendrils. 92
91. With aerial roots (Caution! Poison Ivy)... Rhus, p. 131

92. Tendrils borne on the persistent leaf-base; stems green... Smilax, p. 45
92. Tendrils opposite the leaf-scars. 93

93. Tendrils ending in flat disks... Parthenocissus, p. 147
93. Tendrils not ending in disks. 94

94. Pith somewhat chambered... Ampelopsis, p. 147
94. Pith continuous... Vitis, p. 147

95. Leaf-scars linear or U-shaped. 96
95. Leaf-scars not as above. 97

96. Buds superposed; stem not prickly... Aristolochia, p. 91
96. Buds solitary; stem prickly... Rosa, p. 117

97. Some twigs spinescent... Lycium, p. 183
97. Entirely unarmed. 98

98. Bearing a knob at each angle of the leaf-scar... Wisteria, p. 130
98. Not as above. 99

99. Buds small or sunken. 100
99. Buds medium-sized. 102

100. Stem hairy; stone of fruit crescent-shaped... Cocculus, p. 94
100. Stem essentially glabrous. 101

101. Stone of fruit crescent-shaped... Menispermum, p. 94
101. Stone cup-shaped... Calycocarpum, p. 94

102. Leaf-scars low, bud-scales glabrous... Celastrus, p. 139
102. Leaf-scars raised, bud-scales pubescent... Solanum, p. 183

103. Bearing spines or prickles. 104
103. Without spines or prickles. 116

104. Bearing spines (stiff outgrowths of the twig). 105
104. Bearing prickles (superficial outgrowths). 111

105. Spines representing leaves or stipules. 106
105. Spines borne singly at the side of the buds... Lycium, p. 183
105. Spines representing modified twigs. 107

106. Spines representing modified leaves... Berberis, p. 93
106. Spines representing stipules... Robinia, p. 130

107. Sap milky... Maclura, p. 89
107. Sap not milky. 108

108. Leaf-scars on torn membranes; spines branched... Gleditsia, p. 127
108. Not as above. 109

109. Spines very sharp-pointed... Crataegus, p. 112
109. Spines representing less modified twigs. 110

110. Stipule-scars none... Pyrus, p. 108
110. Stipule-scars present... Prunus, p. 121

111. Leaf-bases persistent, torn at top... Rubus, p. 115
111. Leaf-bases not persistent. 112

112. Leaf-scars irregularly cracked... Robinia, p. 130
112. Not as above. 113

113. Prickles paired at the nodes... Zanthoxylum, p. 130
113. Not as above. 114

114. Leaf-scars nearly encircling the thick stem... Aralia, p. 157
114. Leaf-scars not as above. 115

115. Buds elongated, stalked, pith porous... Ribes, p. 99
115. Buds ovoid, pith continuous... Rosa, p. 117

116. Without leaf-scars, but with persistent leaf-bases. 117
116. With leaf-scars. 119

117. Leaf-scars much raised on a clasping 3-nerved base... Potentilla, p. 115
117. Not as above. 118

118. Pith central in the branches... Rubus, p. 115
118. Pith toward one side of the branches... Tamarix, p. 154

119. Bundle-traces distinct, 3 or more in a line. 120
119. Bundle-trace one, or many traces scattered or nearly confluent in a line. 191

120. Leaf-scars very narrow. 121
120. Leaf-scars broader. 145

121. Leaf-scars straight or U-shaped. 122
121. Leaf-scars horseshoe-shaped, or ring-like, nearly or completely encircling the bud. 140

122. Leaf-scars somewhat torn by the developing buds... Ptelea, p. 131
122. Not as above. 123

123. Stipule-scars encircling the twig... Magnolia, p. 94
123. Stipule-scars smaller or none. 124

124. Leaf-scars half-encircling the twig. 125
124. Leaf-scars shorter. 126

125. Tall shrub, buds gummy... Sorbus, p. 109
125. Small shrub, buds not gummy... Xanthorhiza, p. 93

126. Bud-scale one... Salix, p. 46
126. Bud-scales two or more. 127

127. Pith porous in old twigs... Ribes, p. 99
127. Pith continuous. 128

128. Buds round-ovoid with resinous or fleshy scales... Baccharis, p. 195
128. Not as above. 129

129. Sap milky... Rhus, p. 131
129. Sap not milky. 130

130. Aromatic. 131
130. Not aromatic. 133

131. Bark peeling horizontally around the stem. 132
131. Bark not as above... Lindera, p. 98

132. Bark pleasantly aromatic, with the odor of wintergreen... Betula, p. 68
132. Bark unpleasantly aromatic... Prunus, p. 121

133. Leaf-scars low, straight... Rosa, p. 117
133. Leaf-scars raised, or curved. 134

134. Lateral buds short-ovoid. 135
134. Lateral buds elongated-ovoid. 136

135. Stipule-scars present... Betula, p. 68
135. Stipule-scars absent... Pyrus, p. 108

136. Bud curved, the scales twisted; pith minute, angled... Amelanchier, p. 109
136. Not as above. 137

137. Buds symmetrical; pith minute, flattened or 3-angled... Betula, p. 68
137. Without this combination of characters. 138

138. Buds woolly or gummy... Sorbus, p. 109
138. Buds not as above. 139

139. Buds oblong or elongated-ovoid... Aronia, p. 109
139. Buds short-ovoid... Prunus, p. 121

140. Stipule-scars encircling the twig... Platanus, p. 105
140. Stipule-scars shorter or none. 141

141. Leaf-scars nearly encircling the twig. 142
141. Leaf-scars torn. 144

142. Sap milky... Rhus, p. 131
142. Sap not milky. 143

143. Nodes swollen... Dirca, p. 154
143. Nodes not swollen... Cladrastis, p. 129

144. Aromatic; leaf-scars U-shaped when torn... Ptelea, p. 131
144. Not aromatic; leaf-scars irregularly torn... Robinia, p. 130

145. Pith chambered. 146
145. Pith continuous. 147

146. Twigs rather thick; leaf-scars large... Juglans, p. 61
146. Twigs slender; leaf-scars small... Celtis, p. 85

147. Pith with firmer diaphragms at intervals. 148
147. Pith without firmer diaphragms. 149

148. Buds densely covered with reddish-brown hairs... Asimina, p. 97
148. Not as above... Nyssa, p. 157

149. Buds densely covered with reddish-brown hairs... Asimina, p. 97
149. Not as above. 150

150. Buds in hairy pits... Gymnocladus, p. 127
150. Buds not in hairy pits. 151

151. Pith angled; twigs corky-ridged... Liquidambar, p. 103
151. Not as above. 152

152. Sap milky. 153
152. Sap not milky. 154

153. Bud-scales 2 or 3... Rhus, p. 131
153. Bud-scales several... Morus, p. 87

154. Buds glossy-varnished; pith angled... Liquidambar, p. 103
154. Not as above. 155

155. Lowest scale of lateral buds centrally located over the leaf-scar; pith angled... Populus, p. 55
155. Not as above. 156

156. Low shrub with leaves persistent until late in winter, fern-like... Comptonia, p. 61
156. Not as above. 157

157. With resin glands or blisters in sheltered places. 158
157. Not as above. 159

158. Stipule-scars elongated; resin in blisters... Betula, p. 68
158. Stipule-scars minute or none... Myrica, p. 58

159. Buds stalked. 160
159. Buds not stalked. 164

160. Buds long and spine-like... Fagus, p. 73
160. Not as above. 161

161. Leaf-scars 2-ranked. 162
161. Leaf-scars in more than 2 ranks. 163

162. Buds yellowish... Hamamelis, p. 103
162. Buds black or brown, only the flower-buds stalked... Cercis, p. 129

163. Aromatic; buds greenish-yellow...Lindera, p. 98
163. Not aromatic; buds reddish...Alnus, p. 71

164. Pith flattened or 3-angled. 165
164. Pith round. 167

165. Bud-scales overlapping each other. 166
165. Bud-scales not overlapping...Alnus, p. 71

166. Bud-scales 4-6...Corylus, p. 67
166. Bud-scales usually 2 or 3...Betula, p. 68

167. Twigs with 3 small vertical ridges below the leaf-scars. 168
167. Not as above. 170

168. Leaf-scars 2-ranked...Cercis, p. 129
168. Leaf-scars in more than 2 ranks. 169

169. Buds appressed; bark exfoliating...Physocarpus, p. 105
169. Not as above...Prunus, p. 121

170. Buds long and spine-like; stipule-scars long...Fagus, p. 73
170. Not as above. 171

171. Twigs quite thick; leaf-scars large; buds short...Ailanthus, p. 131
171. Not as above. 172

172. Buds small and appressed. 173
172. Not as above. 176

173. Leaf-scars on raised leaf-cushions. 174
173. Leaf-scars low. 175

174. Bark peeling horizontally, inner bark with a disagreeable odor...Prunus, p. 121
174. Bark peeling vertically; lenticels prominent...Rhamnus, p. 143

175. Lateral buds triangular...Celtis, p. 85
175. Lateral buds minute, not triangular...Cornus, p. 158

176. Pith 5-angled, twigs often warty...Liquidambar, p. 103
176. Not as above. 177

177. Leaf-scars 2-ranked. 178
177. Leaf-scars in more than 2 ranks. 182

178. Bud-scales in 2 ranks...Ulmus, p. 83
178. Bud-scales not in 2 ranks. 179

179. Bud-scales with vertical striations...Ostrya, p. 67
179. Bud-scales not striate. 180

180. Bud-scales 2... Tilia, p. 151
180. Bud-scales several. 181

181. Buds nearly globose, with 4-6 scales... Corylus, p. 67
181. Buds oblong, with about 12 scales... Carpinus, p. 68

182. Exposed bud-scales 2... Cornus, p. 158
182. Exposed bud-scales more than 2, or the buds naked. 183

183. Without stipules or stipule-scars. 184
183. With stipules or stipule-scars. 186

184. Aromatic; twigs green... Sassafras, p. 98
184. Not aromatic. 185

185. Buds woolly or gummy... Sorbus, p. 109
185. Buds not woolly or gummy... Pyrularia, p. 89

186. Stipule-bases present, or stipule-scars short. 187
186. Stipule-scars elongated... Betula, p. 68

187. Stipule-scars or bases on a leaf-cushion. 188
187. Stipule-scars not on a leaf-cushion. 189

188. Inner bark with a disagreeable odor; lenticels horizontally elongated... Prunus, p. 121
188. Inner bark not with a disagreeable odor; lenticels prominent, not horizontally elongated... Rhamnus, p. 143

189. Bundle-traces confluent, or twigs hairy... Rhamnus, p. 143
189. Bundle-traces separate, or twigs glabrous. 190

190. Lenticels horizontally elongated... Prunus, p. 121
190. Lenticels not horizontally elongated... Crataegus, p. 112

191. Stipule-scars nearly or quite encircling the twig. 192
191. Not as above. 195

192. Buds long and pointed. 193
192. Buds not as above. 194

193. Leaf-scars small... Fagus, p. 73
193. Leaf-scars large... Magnolia, p. 94

194. Buds large, flattened... Liriodendron, p. 95
194. Buds short, ovoid... Ulmus, p. 83

195. Bundle-traces many, mostly in 3 groups; leaf-scars lobed. 196
195. Not as above. 197

196. Pith chambered... Juglans, p. 61
196. Pith continuous... Carya, p. 61

197. Bundle-traces many, not in 3 groups. 198
197. Bundle-trace one. 208

198. Leaf-scars large; bundle-traces in an ellipse... Ampelopsis, p. 147
198. Not as above. 199

199. Sap milky. 200
199. Sap not milky. 203

200. With stipule-scars. 201
200. Without stipule-scars... Rhus, p. 131

201. Pith with thin plates at the nodes... Broussonetia, p. 87
201. Not as above. 202

202. Buds ovoid... Morus, p. 87
202. Buds depressed-globose... Maclura, p. 89

203. Pith angular, more or less star-shaped. 204
203. Pith round or nearly so. 205

204. Bud-scales numerous... Quercus, p. 74
204. Bud-scales 2 or 3... Castanea, p. 73

205. Buds evident. 206
205. Buds not evident... Hibiscus, p. 153

206. Buds asymmetrical... Tilia, p. 151
206. Buds nearly symmetrical. 207

207. Scars of 2 kinds, viz., leaf-scars with 3 bundle-traces, and branch-scars with numerous bundles in an ellipse... Pyrularia, p. 89
207. Not as above... Corylus, p. 67

208. Leaf-scars minute, on small ridges; twigs bearing cones persistent in winter... Larix, p. 35
208. Not as above. 209

209. Pith chambered or porous. 210
209. Pith continuous. 212

210. Bud-scales 2, overlapping... Diospyros, p. 178
210. Bud-scales several. 211

211. Buds deltoid... Celtis, p. 85
211. Buds ovoid... Halesia, p. 178

212. Bundle-trace broken into three. 213
212. Bundle-trace unbroken. 215

213. Twigs aromatic, green... Sassafras, p. 98
213. Not as above. 214

214. Buds solitary... Rhamnus, p. 143
214. Buds superposed... Ilex, p. 134

215. Leaf-scars on large leaf-cushions. 217
215. Not as above. 216

216 Twigs prominently angled or grooved... Cytisus, p. 129
216. Not as above. 218

217. Buds ellipsoid, twigs rounded... Potentilla, p. 115
217. Not as above... Spiraea, p. 105

218. Bud-scales 2, bundle-trace semicircular ... Diospyros, p. 178
218. Not as above. 219

219. Leaf-scars about as high as wide. 220
219. Leaf-scars wider than high. 225

220. Twigs aromatic, green... Sassafras, p. 98
220. Not as above. 221

221. Terminal bud not enlarged, or absent. 222
221. Terminal bud larger than the lateral buds. 223

222. Twigs brown or red... Oxydendrum, p. 171
222. Twigs gray... Lyonia, p. 170

223. Outer bud-scales as long as the bud... Clethra, p. 161
223. Outer bud-scales shorter than the bud. 224

224. Bark shredding; capsules bristly... Menziesia, p. 167
224. Not as above... Rhododendron, p. 163

225. Soft-wooded, or aromatic. 226
225. Neither soft-wooded nor aromatic. 227

226. Soft-wooded; not aromatic... Ceanothus, p. 145
226. Aromatic; not soft-wooded... Sassafras, p. 98

227. With stipules or stipule-scars. 228
227. Without stipules or stipule-scars. 229

228. Buds sometimes superposed... Ilex, p. 134
228. Buds not superposed... Spiraea, p. 105

229. Bud-scales 2, twigs glaucous... Nemopanthus, p. 135
229. Not as above. 230

230. Buds ovoid or oblong, 231
230. Buds globose or nearly so. 234

231. Fruit present in winter, of small round capsules... Lyonia, p. 170
231. Fruit absent in winter. 232

232. Twigs scurfy with peltate scales... Elaeagnus, p. 155
232. Not as above. 233

233. Twigs green or red or warty... Vaccinium, p. 175
233. Twigs not green or warty... Gaylussacia, p. 175

234. Fruit absent in winter; twigs green or red or warty... Vaccinium, p. 175
234. Fruit present in winter. 235

235. Tree with oblong capsules... Oxydendrum, p. 171
235. Shrub with globose capsules... Leucothoe, p. 170

DESCRIPTIONS OF GENERA AND SPECIES OF WOODY PLANTS

TAXUS L. (Taxaceae)

SHRUBS with evergreen, linear, mucronate, alternate leaves, the pistillate flower consisting of a single ovule (not in a cone).

1. T. canadensis Marsh. American Yew. A low sprawling shrub, seldom over 1 m. high; leaves 2-ranked, dark green above, yellowish-green beneath, 12-30 mm. long, about 2 mm. wide, persistent on twigs in drying; seed mostly surrounded by a fleshy aril which when mature becomes 6 mm. long, red, juicy, sweet, resembling the fruit of angiosperms, often persistent into winter. Rich woods and thickets, Newfoundland to Manitoba, south to Iowa and the mountains of Kentucky (Fig. 1).

ABIES Mill. (Pinaceae)

Evergreen trees with leaves linear, flat, scattered, sessile, spreading so as to appear 2-ranked, but in reality spirally arranged, usually persistent in drying (hence the young plants quite satisfactory as Christmas trees). Cones erect, cylindric or ovoid, maturing the first year, the scales falling from the persistent axis (hence the ripe cones do not fall from the tree intact, as in most other conifers).

1. A. balsamea (L.) Mill. Balsam Fir. A slender tree 10-25 m. tall, the trunk 1 m. in diameter; bark nearly smooth, but with resin blisters (hence the name "Blister-Pine" commonly used in the Alleghenies); leaves narrowly linear, obtusely pointed, 1-3.2 cm. long, marked with two white lines beneath; cones cylindrical, 6-10 cm. long when mature; bracts obovate, serrulate, tipped with an abrupt slender point, shorter than or about equaling the scales. Woods, Labrador to Alberta, south to Iowa and the mountains of West Virginia (Fig. 2).

TSUGA (Endl.) Carr. (Pinaceae)

Evergreen trees with slender horizontal or drooping branches and flat narrowly linear alternate leaves, spreading and appearing 2-ranked, attached to very short stalks (sterigmata), quickly fall-

Fig. 1. Taxus canadensis

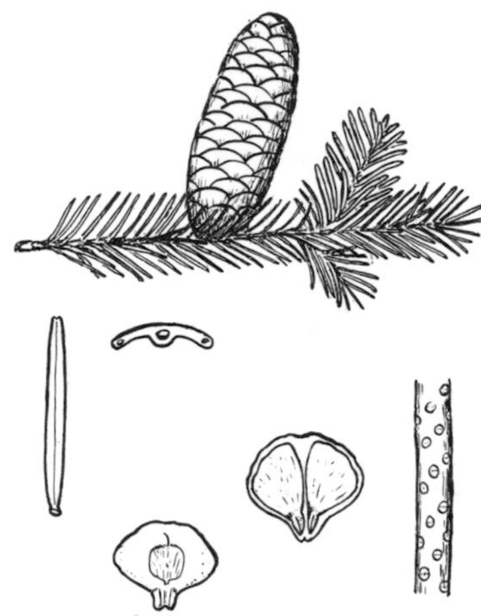

Fig. 2. Abies balsamea

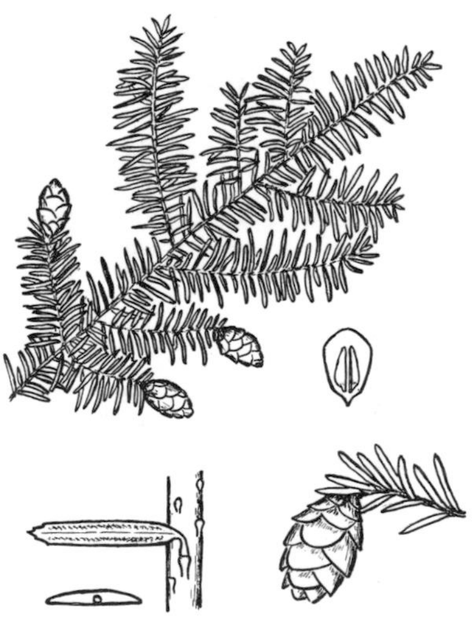

Fig. 3. Tsuga canadensis

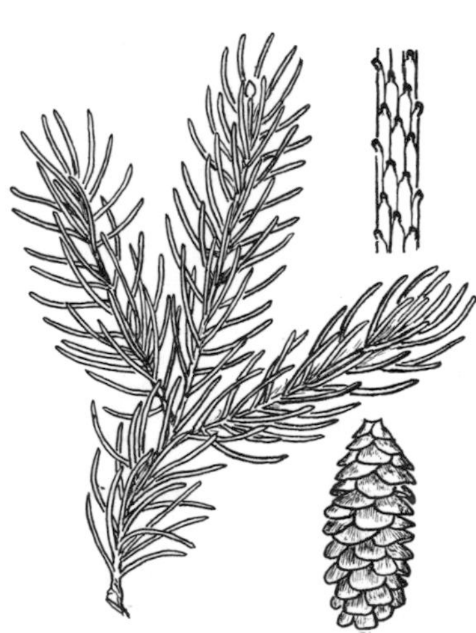

Fig. 4. Picea glauca

ing when dry. Cones small, ovoid or oblong, pendulous, the scales thin, obtuse, persistent.

1. T. canadensis (L.) Carr. Hemlock. A tall tree, 20-35 m. high, 6-12 dm. in diameter, with gracefully spreading spray; bark deeply fissured on old trunks, with prominent rounded ridges; foliage delicate, dense; leaves short-petioled, linear, obtuse, 8-13 mm. long; cones ovoid, 1.5-2.5 cm. long, maturing the first autumn, the scales nearly orbicular. Woods, New Brunswick to Minnesota, south along the mountains to Georgia and Alabama (Fig. 3).

PICEA Dietr. (Pinaceae)

Evergreen conical trees, with linear short more or less 4-sided leaves spreading in all directions, joined at the base to short sterigmata, quickly falling when dry. Cones ovoid or oblong, obtuse, pendulous, the scales numerous, thin, obtuse, persistent.

a. Cones usually less than 5 cm. long
  - b. Twigs glabrous — 1. P. glauca
  - b. Twigs pubescent
    - c. Leaves 12-15 mm. long; Appalachian — 2. P. rubens
    - c. Leaves usually 6-10 mm. long; chiefly Canadian — 3. P. mariana

a. Cones 10-15 cm. long — 4. P. abies

1. P. glauca (Moench.) Voss. White Spruce. A handsome tree to 45 m. high, 4-6 dm. in diameter; bark pale brown; branchlets glabrous; leaves slender, pale or glaucous, 8-18 mm. long; cones subcylindrical, about 5 cm. long, the scales pale, thin, with an entire edge. Rich woods, Labrador to Alaska, south to South Dakota and New York; typically a Canadian tree (Fig. 4).

2. P. rubens Sarg. Red Spruce. (P. rubra Dietr.). A slender tree 20-35 m. high, 6-9 dm. in diameter; bark reddish, roughened by thin irregular brown scales; branchlets pubescent; leaves slender, 12-15 mm. long, somewhat acute; cones maturing the first year, deciduous in autumn or during the winter, elongate-ovoid, 3-4 cm. long, brown, the scales rounded, entire, or slightly erose at the tip. Woods, Prince Edward Island to Ohio, south on the mountains to North Carolina and Tennessee; typically an Appalachian tree (Fig. 5).

3. P. mariana (Mill.) BSP. Black Spruce. Bog Spruce. Usually a small tree less than 30 m. high and 2-3 dm. in diameter;

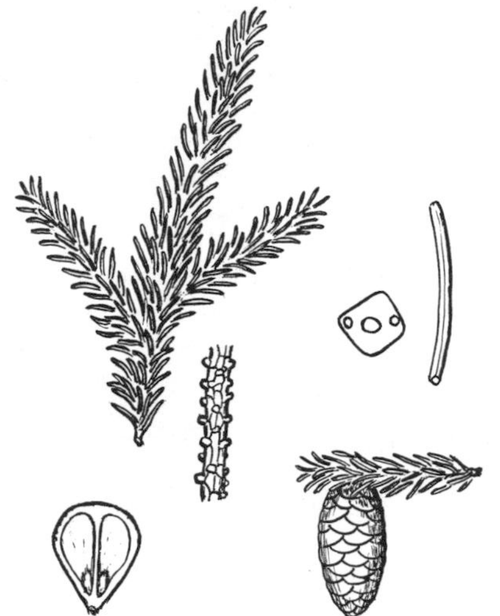

Fig. 5. Picea rubens

Fig. 6. Picea mariana

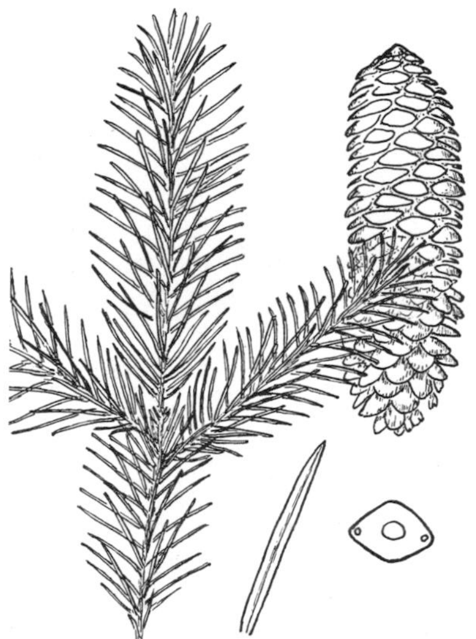

Fig. 7. Picea abies

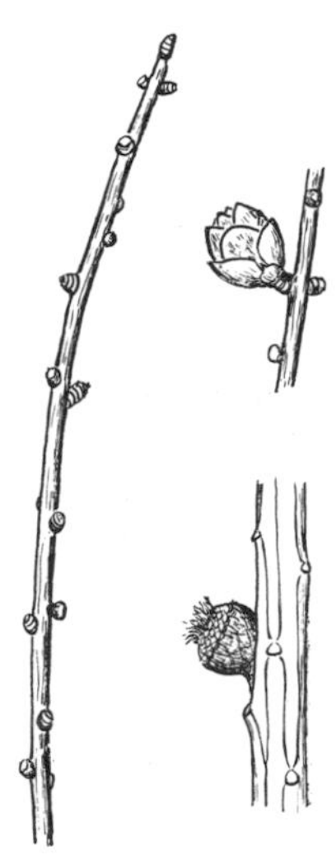

Fig. 8. Larix laricina

bark broken into flaky gray or reddish scales; branchlets pubescent; leaves short and thick, mostly 6-10 mm. long, pale bluish-green with a white bloom; cones short-ovoid or globose, 2-3 cm. long, dull grayish-brown, the scales erose. Cool slopes and bogs, Labrador to Alaska, south to Alberta and Pennsylvania, with scattered colonies in the mountains of Virginia; typically a Canadian tree (Fig. 6).

4. P. abies (L.) Karst. Norway Spruce. (P. excelsa Link). A tree to 50 m. high, with drooping lower branches; bark reddish-brown; buds reddish or light brown, not resinous; branchlets nearly glabrous; leaves slender, sharp-pointed, dark-green, glossy; cones 1-1.5 dm. long. Introduced from Europe as a shade tree, spreading slightly from cultivation in the northern section(Fig. 7).

LARIX Mill. (Pinaceae)

Trees with horizontal or ascending branches and small deciduous leaves, mostly clustered at the ends of short stubby branchlets. Cones ovoid or cylindric, small, erect, the scales thin, persistent.

1. L. laricina (Du Roi) K. Koch. Larch. Tamarack. A slender tree 10-20 m. high, 3-6 dm. in diameter, the branches spreading; bark thin, roughened with small, rounded, red-brown scales; twigs slender, orange, pith minute, brown, roundish; buds solitary, sessile, globose or short-ovoid, with numerous brown scales; leaf-scars minute, alternate, raised on decurrent bases, mostly closely clustered on spurs that elongate very slowly; bundle-trace 1; stipule-scars none; cones less than 2 cm. long, persistent into winter. Labrador to Alaska, south mostly in swamps to Minnesota and the mountains of West Virginia (Fig. 8).

PINUS L. (Pinaceae)

Evergreen trees with 2 kinds of leaves, the primary ones solitary, scalelike, deciduous the first year, the secondary forming the ordinary foliage, narrowly linear, arising in the axils of the primary in bundles (fascicles) of 2 to 5, persistent for several years. Cones requiring 2 years for ripening, those of the first year small and green in winter, maturing the second autumn but often persisting for several years, usually armed with prickles or spines.

a. Leaves 5 in a fascicle — 1. P. strobus

a. Leaves 2 or 3 in a fascicle

b. Leaves 3 in a fascicle — 7. P. rigida

Fig. 9. Pinus strobus

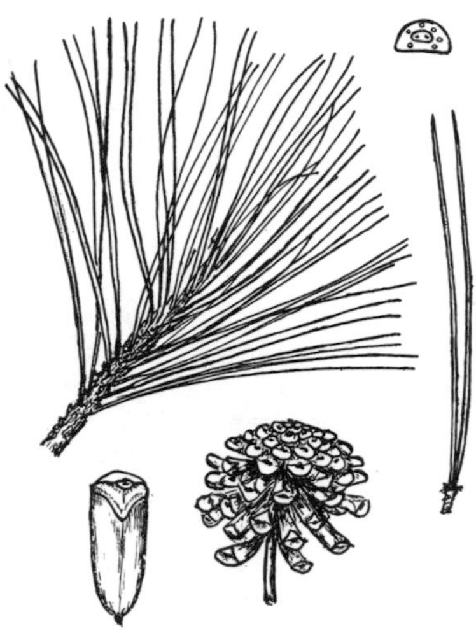

Fig. 10. Pinus resinosa

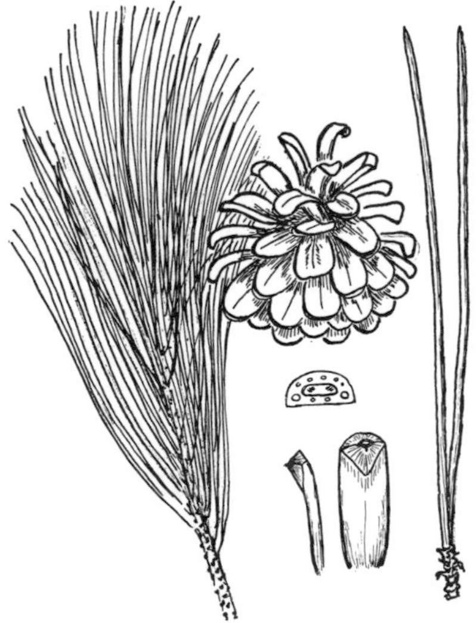

Fig. 11. Pinus nigra

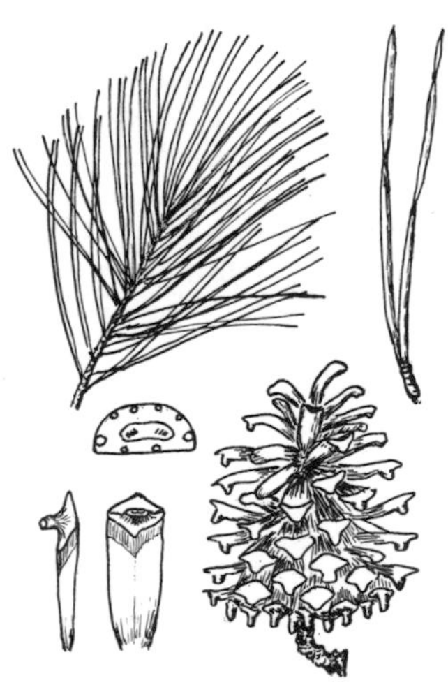

Fig. 12. Pinus sylvestris

b. Leaves 2 in a fascicle, or in 2's and 3's

c. Cone-scales unarmed

d. Leaves 2-4 cm. long; cones often curved — 9. P. banksiana

d. Leaves 3-17 cm. long; cones usually not curved

e. Leaves 7-17 cm. long; cone-scales without a central protuberance — 2. P. resinosa

e. Leaves 3-7 cm. long; cone-scales with a tall central protuberance — 4. P. sylvestris

c. Cone-scales armed

d. Leaves in 2's and (often) 3's on the same tree

e. Leaves 7-13 cm. long; prickle weak — 5. P. echinata

e. Leaves 3-7 cm. long; prickle very strong — 8. P. pungens

d. Leaves 2 in a fascicle

e. Leaves 2-4 cm. long — 9. P. banksiana

e. Leaves 4-17 cm. long

f. Leaves 4-8 cm. long; native tree — 6. P. virginiana

f. Leaves 7-17 cm. long; introduced tree — 3. P. nigra

1. P. strobus L. White Pine. A large forest tree 20-50 m. high, with a trunk diameter of 6-13 dm.; bark on the young branches smooth, on the old trunks divided by shallow grooves into wide flat-topped ridges; leaves in 5's, 7-13 cm. long, slender, marked with white lines, the fascicle-sheaths deciduous; mature cones slender, cylindrical, somewhat curved, 1-1.5 dm. long, the scales unarmed. Rich woods, Newfoundland to Manitoba, south to Iowa and the mountains of Georgia and Tennessee (Fig. 9).

2. P. resinosa Ait. Red Pine. A tall forest tree reaching a maximum height of 50 m. and a diameter of 15 dm.; bark reddish,

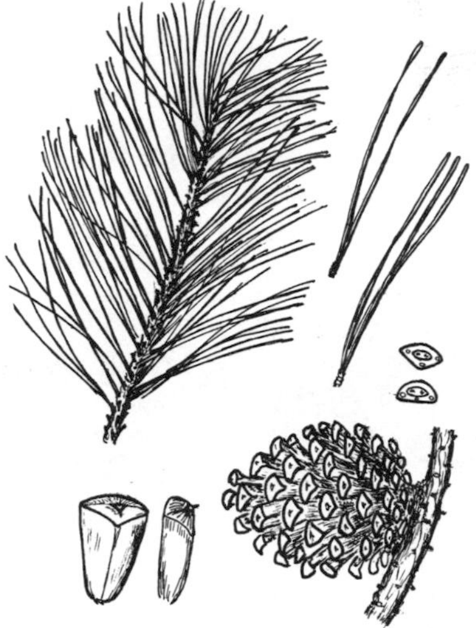

Fig. 13. Pinus echinata

Fig. 14. Pinus virginiana

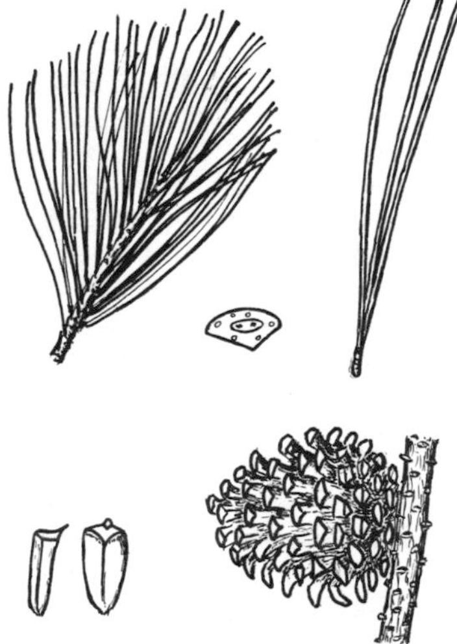

Fig. 15. Pinus rigida

Fig. 16. Pinus pungens

rather smooth; leaves 2 in a fascicle, dark green, 9-16 cm. long; mature cones ovoid-conical, 5 cm. long; scales unarmed. Dry woods, Newfoundland to Manitoba, south to Minnesota and the mountains of West Virginia; most common about the Great Lakes and in New England (Fig. 10).

3. P. nigra Arnold. Austrian Pine. A tree 30-50 m. high; bark dark brown to black; buds silvery or light brown; branchlets usually light brown; leaves stiff, 9-16 cm. long, dark green; cones 5-8 cm. long, the scales with a short prickle on the umbo. Introduced from Europe and spreading slightly from cultivation (Fig. 11.)

4. P. sylvestris L. Scotch Pine. A timber tree up to 40 m. high; bark gray; leaves in 2's, bluish or grayish-green, 3-7 cm. long; cones 3-6 cm. long, slender, conic, reflexed, the scales thickened, rhombic, with a central protuberance but not prickly. Introduced from Europe, much cultivated, and naturalized (Fig. 12).

5. P. echinata Mill. Shortleaf Pine. Yellow Pine. A straight tree 15-45 m. high, with a trunk diameter of 6-10 dm.; bark broken into large more or less rectangular plates; leaves mostly in 2's, but also in 3's on the same tree, 7-13 cm. long, slender, flexible; mature cones ovoid, about 5 cm. long; scales armed with a minute weak prickle. Dry or sandy soil, Florida to Texas, north to New York, Ohio, and Oklahoma (Fig. 13).

6. P. virginiana Mill. Scrub Pine. Virginia Pine. Jersey Pine. A straggling tree 5-20 m. high, trunk diameter 3-6 dm.; branches spreading or drooping; bark with shallow fissures and dark brown loose scales; leaves in 2's, 4-8 cm. long, twisted, gray-green; mature cones few, oblong-conic, 3-7 cm. long;scales with a straight or curved slender prickle. Barrens and sterile soil, Georgia to Arkansas, north to Virginia, New Jersey, New York and Indiana (Fig. 14).

7. P. rigida Mill. Pitch Pine. A tree 10-25 m. high, with very rough dark bark and hard resinous ("pitchy") wood; leaves in 3's, 5-12 cm. long, rigid, dark green; mature cones ovoid-conical or ovoid, 3-9 cm. long, often clustered; scales thickened at the apex, bearing a short thick curved prickle. Sandy or barren soil, Maine to Ontario, south to Ohio and the mountains of Georgia and Tennessee (Fig. 15).

8. P. pungens Lamb. Table Mountain Pine. A rather small tree, 6-20 m. high, and with trunk diameter of 10 dm. ; bark broken by fissures into irregular red-brown plates, leaves in 2's

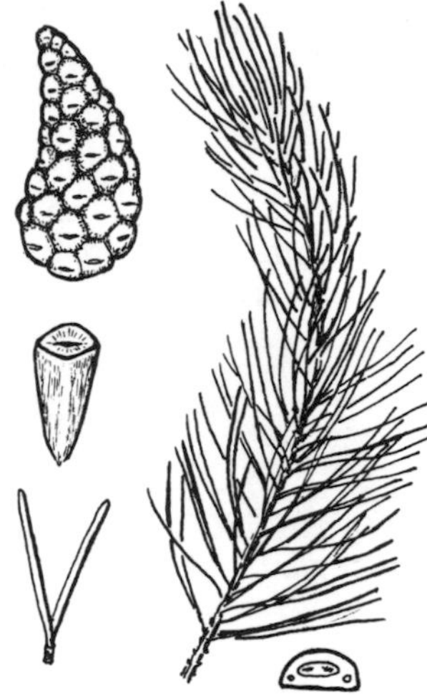

Fig. 17. Pinus banksiana

Fig. 18. Taxodium distichum

Fig. 19. Thuja occidentalis

Fig. 20. Chamaecyparis thyoides

(sometimes also in 3's), stiff, 3-6 cm. long, blue-green; mature cones clustered, ovoid, 5-10 cm. long, hanging on for many years; scales very thick and woody, armed with a strong hooked spine 4-5 mm. long. Uplands, New Jersey and Pennsylvania to Georgia and Tennessee (Fig. 16).

9. P. banksiana Lamb. Jack Pine. A low tree usually 5-20 m. high, 3-4 dm. in diameter; bark thin, brown, reddish or gray, irregularly broken into scaly ridges; leaves in 2's, short and thick, 2-4 cm. long, divergent; cones conical, oblong, usually curved, 3-5 cm. long, the scales smooth or with a minute prickle. Barren or rocky soil, Quebec to Mackenzie, south to Alberta, Michigan and New York (Fig. 17).

TAXODIUM Richard (Taxodiaceae)

Deciduous trees with shreddy bark, often buttressed when large and in very wet places surrounded by large conical "knees" growing upwards from the roots. Cones small, ellipsoid, with thickened scales.

1. T. distichum (L.) Richard. Bald-Cypress. Tree to 40 m. high, 9-15 dm. in diameter, the base conical or abruptly enlarged, more or less ridged; bark fibrous or scaly, thin, reddish-brown; twigs slender; pith minute, brown, roundish; buds sessile, minute, subglobose; leaf-and stipule-scars lacking. Swamps, Florida to Texas, north to New Jersey, Kentucky and Oklahoma (Fig. 18).

THUJA L. (Cupressaceae)

Evergreen trees or shrubs with flattened sprays, the leaves very small, scale-like, appressed, overlapping each other, opposite, 4-ranked. Cones ovoid or oblong, mostly spreading, the 6-10 scales opposite, dry, spreading when mature.

1. T. occidentalis L. Arborvitae. A conical tree 10-35 m. high, 3-6 dm. in diameter, with pale shreddy bark and light, soft, but durable wood; leaves 2-3 mm. broad, of two sorts, the two lateral rows keeled, the two other rows flat, causing the twig to appear much flattened; cones oblong, 8-12 mm. long, reddish-brown, maturing in early autumn, persisting through the following winter; seeds broadly winged all around. Swamps and rocky banks, Quebec to Saskatchewan, south chiefly on limestone outcrops to Minnesota, Ohio, and the mountains of North Carolina and Tennessee (Fig. 19).

Fig. 21. Juniperus communis

Fig. 22. Juniperus horizontalis

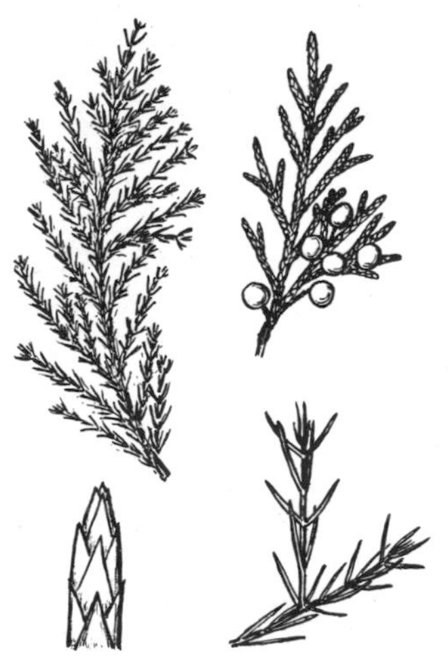

Fig. 23. Juniperus virginiana

Fig. 24. Smilax rotundifolia

CHAMAECYPARIS Spach (Cupressaceae)

Strong-scented evergreen trees with very small scale-like closely appressed overlapping leaves, 2-ranked branchlets, and subglobose cones with thick peltate scales.

1. C. thyoides (L.) BSP. White Cedar. A tree to 25 m. high, 3-4 dm. in diameter; bark thin, ashy-gray to reddish-brown; leaves minute, pale, often bearing a small gland, closely imbricated in 4 rows; cones small, 6-9 mm. in diameter, with about 3 pairs of scales. Swamps, in the coastal plain, from Mississippi and Florida north to Maine (Fig. 20).

JUNIPERUS L. (Cupressaceae)

Evergreen trees or shrubs with opposite or whorled scale-like sessile leaves, commonly of 2 sorts, those of the young branches linear, often spreading, on the older branches scale-like, closely appressed and overlapping. Cones of 3-6 fleshy coalescent scales, when mature forming a sort of berry, light blue and glaucous.

a. Leaves in whorls of 3 — 1. J. communis
a. Leaves mostly opposite
  b. Prostrate shrub; cones 6-10 mm. in diameter — 2. J. horizontalis
  b. Upright shrub or tree; cones 5-6 mm. in diameter — 3. J. virginiana

1. J. communis L. Common Juniper. A low decumbent shrub (in the var. depressa Pursh) or small tree, 2-10 m. high, with pyramidal or columnar form; bark dark reddish-brown, scaly; leaves thin, straight, 12-21 mm. long, widely spreading, grayish above, sharp-pointed; cones subglobose, 5-10 mm. in diameter. Dry soil, pastures, etc., Greenland to Alaska, south to California, Wyoming, Ohio, and the mountains of Georgia (Fig. 21).

2. J. horizontalis Moench. Creeping Savin. Creeping Juniper. A procumbent, prostrate or creeping shrub; leaves scale-like, sharp-pointed; cones 6-10 mm. in diameter, on a short peduncle. Rocky or sandy banks and bogs, Newfoundland to Alaska, south to Wyoming, Illinois, and New York (Fig. 22).

3. J. virginiana L. Red Cedar. A tree 15-25 m. high (or sometimes only a small tree or shrub), pyramidal in form; bark thin, peeling off in long strips, reddish-brown; leaves mostly opposite, those on the young twigs subulate, spiny-tipped, 4-8 mm.

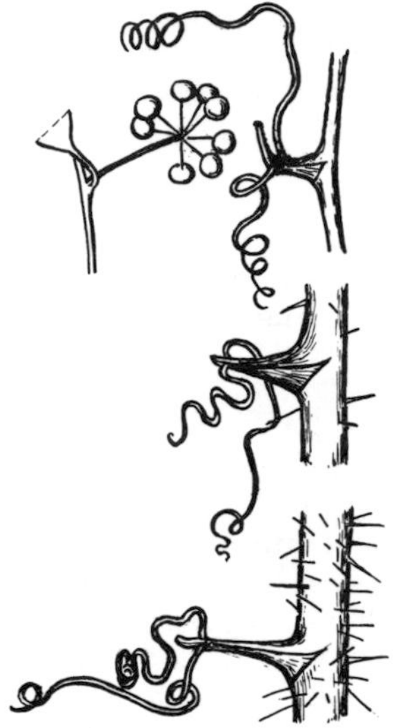

Fig. 25. Smilax hispida

Fig. 26. Smilax glauca

Fig. 27. Smilax laurifolia

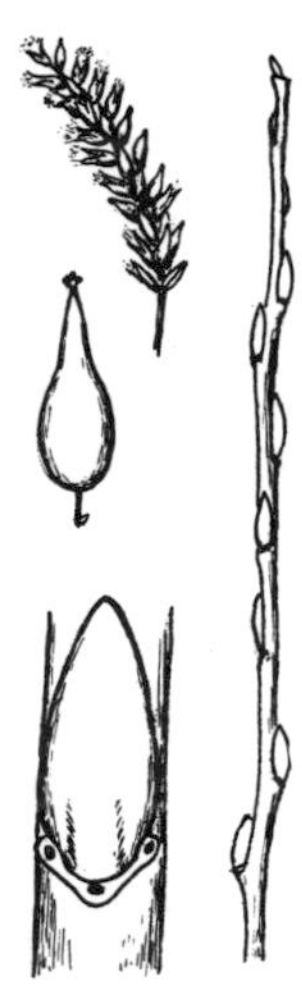

Fig. 28. Salix nigra

long, those of the older branches scale-like, acute or subacute, closely appressed and overlapping, 4-ranked, causing the twigs to appear quadrangular; cones ("berries") maturing in autumn, light blue, glaucous, about 6 mm. in diameter. Dry woods and barrens, often on limestone outcrops, Florida to Texas, north to New England and Missouri (Fig. 23).

SMILAX L. (Liliaceae)

Woody (sometimes herbaceous) plants climbing by tendrils which are borne in pairs on the petiole of the leaves, and may represent modified stipules. Stems terete or angled, usually bearing strong large prickles; pith lacking, the vascular bundles scattered throughout, as is customary in monocots. Buds moderate, 3-sided, pointed, divergent, with a single exposed scale. Leaves deciduous (evergreen in warm climates) tearing away above the broadened base of the petiole and hence leaving no definite scar. Fruit a berry, often persistent into winter.

a. Leaves mostly deciduous

  b. Stems not glaucous

    c. Prickles widened or flattened at the base ..... 1. S. rotundifolia

    c. Prickles needle-like, black ..... 2. S. hispida

  b. Stems glaucous ..... 3. S. glauca

a. Leaves evergreen ..... 4. S. laurifolia

1. _S. rotundifolia_ L. _Common Greenbrier_. Stem glabrous, nearly terete, the branches and young shoots usually sharply 4-angled; prickles scattered, stout, straight or a little curved; berries black, 6 mm. in diameter, ripe in October and November, persistent into winter. Thickets and woods, Florida to Texas, north to Nova Scotia, Ontario, and Oklahoma (Fig. 24).

2. _S. hispida_ Muhl. _Hispid Greenbrier._ (_S. tamnoides_ L. var. _hispida_ (Muhl.) Fernald). Stem climbing in thickets, glabrous, terete, the lower parts thickly hispid with numerous straight slender black prickles, the younger branches unarmed or nearly so; branches somewhat angled; berries bluish-black, 6 mm. in diameter. Rich woods, often on limestone outcrops; New York and South Dakota, south to Georgia and Texas (Fig. 25).

3. _S. glauca_ Walt. _Glaucous Greenbrier._ Stem glabrous, terete, glaucous, armed with rather stout numerous or scattered

prickles;branchlets somewhat angled;berries bluish-black,6 mm. in diameter. Sandy thickets and fields, Florida and Texas, north to New England, West Virginia, and Missouri (Fig. 26).

4. S. laurifolia L. Laurel-leaf Greenbrier. Blaspheme-Vine. Evergreen high-climbing vine, with knotty somewhat woody rhizomes and strong terete stems armed with rigid prickles; tendrils intermittent; leaves coriaceous, 0.6-2 dm. long, 1-7.5 cm. broad, the middle vein much more prominent than the lateral ones; berries black. Swamps and moist soil, Florida to Texas, north to New Jersey and Tennessee; also in the West Indies(Fig. 27).

SALIX L.* (Salicaceae)

Deciduous trees and shrubs, with wood of light weight. Tree trunks 1 or 2—4 together, often leaning; shrubs often clumpy and sometimes forming thickets by the growth of underground runners. Twigs mostly slender but sometimes stoutish (S. discolor, S. caprea), terete. Bark on young twigs usually smooth; green, yellow, brown, purplish, or black, sometimes shining; glabrous, glabrate, puberulent, gray-pubescent, or white-tomentose, varying with species, age of twigs, and season. Pith small, round, continuous, white. Buds small to large, 1-2 mm. to 10 mm. or more long, oblong to ovate, appressed, solitary (or sometimes multiple), with a single exposed scale, colored and clothed as the twigs; terminal bud lacking. Leaf-scars alternate or rarely subopposite or opposite (S. purpurea), shallow, narrow, curved; bundle-traces 3. Stipule-scars tiny to obvious, sometimes absent. Aments of precocious species sometimes appearing in February or March.

- a. Trees, twigs mostly slender, yellowish to brown or black; buds small to midsized (except S. fragilis), 1-5 mm. long
  - b. Twigs not markedly pendulous (weeping)
    - c. Twigs yellow or yellowish, glabrous — 4. S. amygdaloides
    - c. Twigs yellowish or greenish (seasonal) to dark brown or black; native trees
      - d. Small tree or shrub; bark gray; twigs not brittle; buds small, 1-3 mm. long — 3. S. caroliniana
      - d. Large tree; bark dark brown to black; twigs brittle at base; buds small to midsized, 2-5 mm. long, reddish-brown. — 5. S. nigra

*Text contributed by Carleton R. Ball

c. Twigs greenish-yellow (young) to reddish-brown (older); introduced trees

d. Bark gray, thick, rough; twigs very brittle at base; buds midsized to large, 3-7 (-10) mm. long — 6. S. fragilis

d. Bark gray-brown, rough, ridged; twigs not brittle; buds small to midsized, 1-3 (-5) mm. long — 7. S. alba

b. Twigs long, pendulous (weeping) — 8. S. babylonica

a. Shrubs, or occasionally small trees (as also is S. caroliniana above)

b. Buds alternate

c. Thicket-forming, stoloniferous shrub, 2-3m. high; twigs slender; buds small, 2-4 mm. long — 9. S. interior

c. Not thicket-forming or stoloniferous shrubs

d. Buds small to midsized, 2-6 or rarely 7 mm. long (4-7 (-10) mm. in S. discolor)

e. Older twigs and buds glabrous (except in hairy varieties, as noted)

f. Twigs midsized to stout, 2-6 (-7) mm. wide at base

g. Seasonal twigs yellowish-brown,

older reddish-brown or darker, shining, not clearly furrowed, the larger somewhat grayish at base

h. Bark brown, twigs longish (hairy in var. intonsa) — 1. S. lucida

h. Bark olive-green, shining, twigs shorter — 2. S. serissima

g. Seasonal twigs yellowish-brown, older dark brown or blackish, shallowly straight-furrowed, some partly shining, many short and divaricate — 11. S. pyrifolia

f. Twigs mostly slender to occasionally midsized, 1-4 (-5)mm. wide at base

g. Bark gray; twigs slender, leafy, young yellowish and occasionally puberulent, older dark brown or reddish-brown, glabrous, seldom shining — 13. S. petiolaris

g. Bark brown; twigs slender or sometimes midsized, longish, young yellowish-brown to dark brown, puberulent, older dark brown to black, glabrous or sometimes micro-puberulent (plants with older twigs pubescent probably represent S. subsericea) — 14. S. sericea

f. Twigs stoutish, reddish or reddish-purple to

mostly dark brown, youngest thinly pubescent (more densely so in var. latifolia), older glabrous; buds large, 4-7 (-10)mm. long 16. S. discolor

e. Twigs and buds more or less hairy, not shining.

f. Tall shrubs to 5 or 6 m. high; twigs and buds mostly midsized

g. Bark brown; twigs slender to midsized, longish, 2-5 mm. in basal diameter, yellow to yellowish-brown 10. S. rigida

g. Bark brown;twigs midsized to stoutish, 2-5 (-7)mm. in basal diameter, short to long, yellow or yellowish-brown to dark brown(densely white-tomentose on the common var. albovestita, with buds to 8 mm. long) 12. S. glaucophylloides

f. Shrub or small tree, 2 to 6 (-9)m. high; bark grayish, rough, scaly;twigs 1-3(-4) mm. at base, reddish to dark brown, often divaricate, the shorter often crooked; buds midsized 19. S. bebbiana

f. Low, many-stemmed shrub to 1 m. high; twigs slender, 1-3 mm.

in diameter; yellowish-brown to brown; buds small, 1-3 (-4)mm. long 18. S. tristis

d. Buds midsized to large, 4-8 (-10) mm. long, pubescent

e. Introduced shrub or small tree, bark gray; 1-2 year twigs yellowish, older dark brown, stoutish 15. S. caprea

e. Native shrubs or small trees; bark brown; twigs mostly reddish-brown to dark brown

f. Few-stemmed shrub or small tree, 2-7 m. high; bark brown; younger twigs often pubescent (more densely so in var. latifolia), older glabrous, stoutish 16. S. discolor

f. Sprangly shrub 1-3 m. high; bark brown; twigs midsized, pubescent, younger yellowish 17. S. humilis

b. Buds opposite, subopposite, or alternate (on same twig), midsized to large, 3-8(-9)mm. long, glabrous (introduced shrub) 20. S. purpurea

1. S. lucida Muhl. Shining Willow. Shrub or small tree 3-6 (-9)m. high, the trunk to 1.2 dm. in diameter; bark brown, glabrous; twigs mostly midsized (2-5 mm. in diameter), chestnut-brown to reddish-brown or darker, glabrous or young twigs more or less hairy (var. intonsa Fernald); buds midsized, 4-6 mm. long, colored and clothed as the twigs; leaf-scars lunate; bundle-traces conspicuous. Low grounds, Delaware to Iowa and North Dakota, north to Newfoundland and west to Manitoba.

2. S. serissima (Bailey) Fernald. Autumn Willow. Shrub 1-4 m. high, branches with olive-brown, lustrous bark, twigs slender, yellowish-brown to brown, shining, glabrous; buds mis-

sized, lance-oblong, 5-7 mm. long, chestnut-brown to olive or reddish-brown, glabrous, shining. Swampy and boggy ground, rather than along streams, from New Jersey to Ohio and Indiana, Minnesota, Montana, north to Newfoundland, James Bay, Saskatchewan, and Alberta.

3. S. caroliniana Michx. Carolina Willow. Ward's Willow. Shrub or small tree to 19 m. high in our area, larger westward; bark gray, deeply checkered; twigs slender, the seasonal 0.5-1 mm., older to 4 mm. in diameter, yellowish, becoming brown to black with age, the youngest pubescent to glabrous; buds very small, 1-3 mm. long, colored and clothed as the twigs. Stream banks and low woods, Florida, Texas, north to Maryland, Pennsylvania, Illinois and Kansas: also in Cuba.

4. S. amygdaloides Andersson. Peachleaf Willow. Shrub or small tree with 1-3 often leaning trunks to 12 m. high and to 4 dm. in diameter; bark brown, scaly or fissured; twigs slender, 1-3 mm. in diameter, yellow, glabrous; buds small, 2-4 mm. long, yellow or yellowish-brown. Alluvial soils, usually near water, Pennsylvania to Texas and Arizona, north to Quebec and British Columbia.

5. S. nigra Marsh. Black Willow. A shrub or tree to 30 m. high, with 1-4 often leaning trunks to 5 dm. in diameter, the largest of the native willows; bark flaky, dark brown to black; twigs slender, 1-3 mm. in diameter, greenish to dark brown, glabrescent, somewhat brittle at base; buds small, 2-4 mm. long, reddish-brown. Abundant in alluvial soils along streams and in low woods, Florida to Texas, north to New Brunswick, Nebraska, and Minnesota (Fig. 28).

6. S. fragilis L. Brittle Willow. Crack Willow. English Willow. Tree to 20 or 25 m. high, sometimes with 2 or 3 trunks, 1-2 m. in diameter; bark thick, rough, gray; twigs slender, greenish-yellow to reddish-brown, pubescent to glabrous, somewhat lustrous, very brittle at base (hence the Latin and English names); buds midsized, 3-7 mm. or occasionally 10 mm. long, brown. Widely introduced from Europe in colonial days, for sentiment and for charcoal to use in making gunpowder; sparingly escaped along watercourses near towns and farmsteads. Extensively hybridized with S. alba both in Europe and America.

7. S. alba L. White Willow. Cricket-bat Willow. Large tree to 20 m. or more in height, and to 1 m. or more in diameter, often with two or three large trunks above the base; bark rough, coarsely ridged, gray to brownish; seasonal twigs greenish-yellow,

pubescent; older twigs reddish-brown, glabrous; buds very small, 1-3 (-5) mm. long, colored as twigs. Var. vitellina (L.) Koch has young twigs yellow, but is very rare in America where most of the collection so named really are hybrids of S. fragilis and S. alba. Widely introduced from Europe in early days for ornament, basketry, poles, charcoal for making gunpowder and medicinal uses; escaped to a considerable extent in some places.

8. S. babylonica L. Weeping Willow. A tree to 12 m. or more high and 1 m. in diameter; bark grayish-brown; twigs very slender, elongated, pendulous, the younger yellow, gradually becoming brownish to brown, tough, puberulent to glabrous; buds small, 1-4 mm. long, yellowish to brown, late-developing. Widely introduced from Europe, especially for cemeteries and as an ornamental, sparingly escaped. Not native to the Babylon area but the Latin name sentimentally based on Psalm 137 because long grown in Europe as a "mourning tree".

9. S. interior Rowlee. Sandbar Willow. Longleaf Willow. A thicket-forming, stoloniferous shrub 2-5 m. high; stems 3-6 or rarely 10 cm. in diameter; bark grayish; twigs slender, red or reddish-brown, glabrescent-glabrous; buds small, 2-4 mm. long, late in developing. Common on alluvial soils, mudbars, sandbars, and beaches. Pennsylvania, Virginia, West Virginia, and Kentucky, thence south to Louisiana and Texas, west to the western edge of the Great Plains in the United States and southern Canada, and northwest to James Bay, Mackenzie, Yukon and Alaska.

10. S. rigida Muhl. Cordate Willow. Heartleaf Willow. Several-stemmed shrub, 2-4 (-6) m. high; twigs slender to mid-sized, 1-5 mm. in diameter, yellowish to becoming dark brown, the younger pubescent; buds small, 2-5 mm. long, colored and clothed as the twigs. Common along stream banks and ditches and in mostly low grounds, Delaware to Virginia, Kentucky, Missouri, and Nebraska, north to Nova Scotia and west to James Bay, Saskatchewan, and northern Montana.

11. S. pyrifolia Andersson. Balsam Willow. Shrub 1-5 m. high, with usually clustered stem but occasionally tree-like; bark grayish, smooth; twigs yellowish (young) to reddish-brown or dark olive brown, glabrous, shining; buds 2-6 mm. long, stout, scarcely pointed, colored as twigs. Moist to wet or swampy ground, Nova Scotia and northern New England to Minnesota and Alberta, north to Labrador and northern British Columbia.

12. S. glaucophylloides Fernald. Dune Willow. Usually many-stemmed shrub 2-4 or 5 m. tall, occasionally tree-like;

twigs usually rather short and stout, yellowish to chestnut-brown and dark-brown, younger somewhat pubescent (or white-tomentose in var. albovestita); buds midsized, 2-5 (-8)mm. long, stoutish, colored and clothed as the twigs. Sand dunes and sandy shores, calcareous slopes, and sometimes in swamps, Maine to Indiana and Wisconsin, north to Newfoundland and Hudson Bay.

13. S. petiolaris J. E. Smith. Slender Willow. Clumpy shrub to a few-stemmed tree, 2-5 or rarely 6-7 m. high; bark gray; twigs slender, leafy, yellowish or yellowish-green and puberulent to dark brown and glabrous (reddish-brown in dry areas); buds small, 2-5 or rarely 6 mm. long, acute to obtuse, colored as the twigs. Moist meadows, streams, and lake shores, New Jersey to Nebraska, Colorado, and Montana, north to New Brunswick, James Bay, and Alberta.

14. S. sericea Marsh. Silky Willow. Shrub with clustered stems or small tree, 2-6 or 8 m. high; bark gray; twigs slender, light brown to mostly dark brown, younger somewhat pubescent to puberulent, older glabrate-glabrous (plants with older twigs pubescent probably represent S. subsericea (Andersson) Schneider, which occurs from Massachusetts to Nebraska and north to Nova Scotia and Alberta); bud-scales small, 2-5 or rarely 6 mm. long, colored and clothed as the twigs. In moist, rocky to gravelly ground, often near or in running water, Georgia to Arkansas, north to Nova Scotia, New Brunswick, Michigan, and Iowa.

15. S. caprea L. Goat Willow. Florist's Willow. Tall shrub or small tree, sometimes with a single trunk, 2-3 m. high; bark gray; twigs stout, mostly 3-5 (-6) mm. in diameter, yellowish-brown to dark brown, pubescent to glabrescent; buds stout at maturity, midsized to large, 4-9 mm. long, younger acute, older obtuse, colored and clothed as the twigs. Introduced from Europe, where it has many and varied uses. Sparingly cultivated in America, chiefly as a "pussy" willow, and used extensively by florists for early spring decorations, as are the introduced S. cinerea L. and the native S. discolor Muhl. Escaped around old nurseries. S. cinerea is closely related but twigs and buds are more densely pubescent.

16. S. discolor Muhl. Pussy Willow. Glaucous Willow. Few-stemmed shrub or small tree, 2-5 or rarely 7.5 m. high; bark thin, usually smooth, reddish-brown; twigs stout, 3-6 mm. in basal diameter, reddish or reddish-purple to mostly dark brown, glabrous or youngest thinly pubescent (more densely so in var. latifolia Andersson);buds large, 4-9 mm. long, acute to obtuse, colored and clothed as the twigs. Common in swamps and moist

lowlands, Delaware to Kentucky, Missouri, South Dakota, and Montana, north to Newfoundland, James Bay, and British Columbia.

17. S. humilis Marsh. Upland Willow. Prairie Willow. Sprangly shrub with several spreading stems, 1-3 m. high; twigs midsized to stoutish, 2-4 mm. in basal diameter, greenish-yellow or yellowish-brown to orange-red or reddish-brown to dark brown, pubescent to puberulent to glabrate; buds midsized to large, 5-8 mm. long, colored and clothed as the twigs. Scattered but common on upland areas in open woodlands, dry barrens, rocky bluffs, sandy ground, and prairies; occasionally in swampy areas, Florida to eastern Texas, north to Newfoundland, Quebec, North Dakota, and Saskatchewan.

18. S. tristis Ait. Dwarf Upland Willow. Dwarf Prairie Willow. Low, many-stemmed shrub, 0.4-1 m. high; twigs slender, leafy, 1-3 mm. in basal diameter, yellowish-brown to brown, pubescent; buds small, 1-3 mm. long, pubescent. Similar to S. humilis but smaller in every way. Occasional on dry and often sandy uplands, roadsides, thicket-borders, mountain balds, etc.; irregularly distributed from Florida to Oklahoma, north to Quebec and Saskatchewan.

19. S. bebbiana Sarg. Bebb Willow. Beaked Willow. Few-stemmed shrub or small tree (sometimes single trunk), 2-6 or rarely 9 m. high; bark grayish, rough and scaly; twigs slender to midsized, 1-3 mm. in basal diameter, often divaricate, the shorter often crooked, reddish to brown or dark brown, sometimes orange-red or purplish, younger pubescent, older less pubescent to glabrous; buds small, 3-5 mm. long, colored and clothed as the twigs; the projecting petiole-bud scars often conspicuous on basal portion of older twigs. Common on moist to wet or somewhat drier ground from New Jersey to South Dakota, southwest to New Mexico and California, north to Newfoundland and Alaska.

20. S. purpurea L. Purple Osier. Bitter Willow. Many-stemmed shrub 1-2.5 m. tall or taller; bark smooth, very bitter; twigs slender, flexible, yellow to yellowish-brown or reddish-brown, rarely purplish, glabrous, sometimes shining; buds remarkable for often being opposite or sub-opposite as well as alternate, midsized, 3-8 mm. long, acute to obtuse, yellow-brown to reddish-brown, glabrous. Widely introduced in colonial times and cultivated for basket-making, the staminate for ornament; sparingly escaped in the northeastern United States.

POPULUS L.

Deciduous trees. Bark at first smooth, green, whitish, or orange, becoming gray and fissured on older stems. Twigs moderate, terete or somewhat angled; pith small, 6-angled, continuous or nearly so, brown. Buds moderate, ovoid or elongated, sessile, solitary, with several exposed scales, the lowest placed centrally over the leaf-scar. Leaf-scars alternate, broadly crescent-shaped to triangular, somewhat 3-lobed, large; bundle-traces 3, or sometimes compound.

a. Lateral buds with more than 4 exposed scales	
b. Twigs and buds shiny or varnished, mostly glabrous	1. P. tremuloides
b. Twigs and buds gray, dull, buds silky or tomentose	
c. Twigs glabrous	2. P. grandidentata
c. Twigs hairy or tomentose	3. P. alba
a. Lateral buds with about 4 exposed scales	
b. Terminal buds mostly 12 mm. or more long.	
c. Twigs reddish-brown;buds quite resinous, very fragrant when crushed	
d. Native tree of Canada and northern United States	6. P. balsamifera
d. Cultivated tree of horticultural origin	7. P. gileadensis
c. Twigs yellow to yellowish brown;buds slightly resinous	
d. Buds widest at the base tapering toward the apex; planted tree	8. P. eugenei

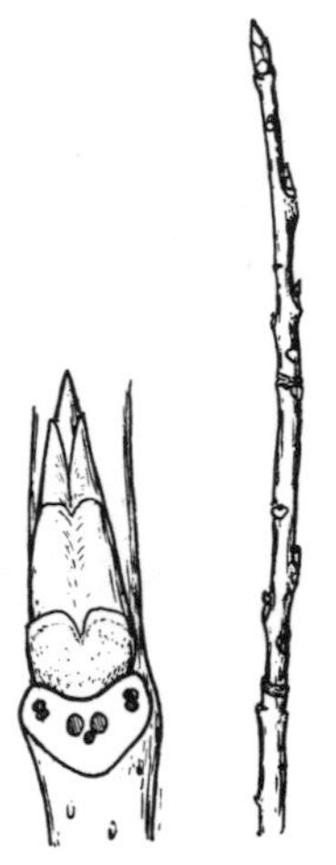

Fig. 29. Populus tremuloides

Fig. 30. Populus grandidentata

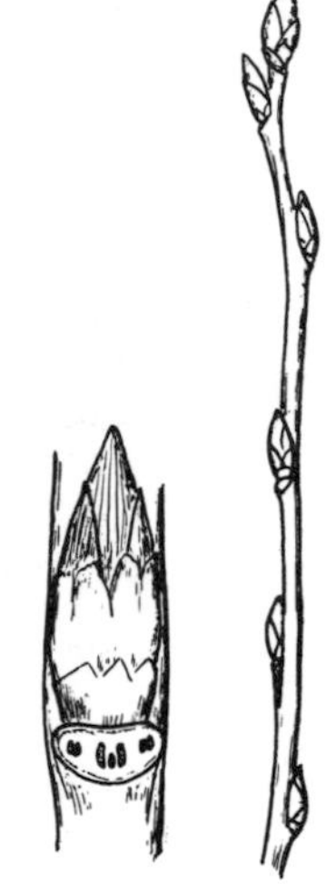

Fig. 31. Populus alba

Fig. 32. Populus deltoides

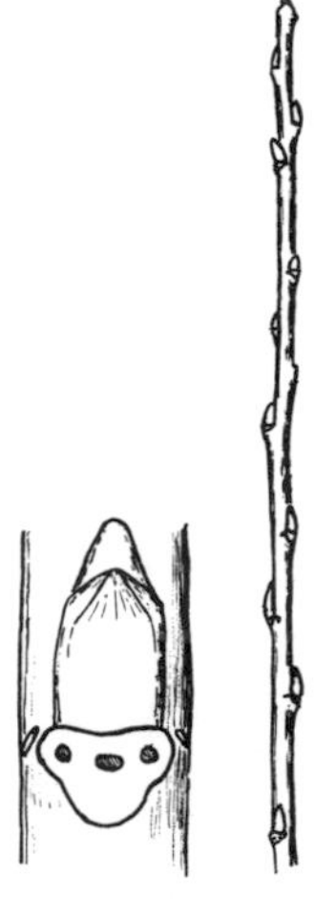

Fig. 33. Populus nigra italica

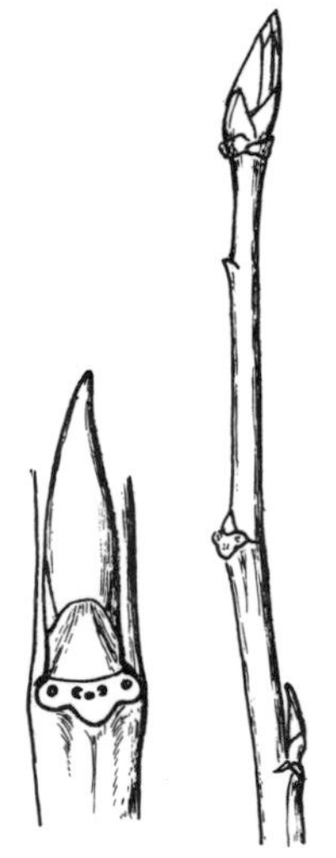

Fig. 34. Populus gileadensis

d. Buds fusiform (widest near the middle, tapering both ways); native tree	4. P. deltoides
b. Terminal buds mostly less than 12 mm. long	
c. Buds short and broad; native swamp tree	9. P. heterophylla
c. Buds slender, tapering; form of tree tall and columnar; introduced	5. P. nigra, var. italica

1. *P. tremuloides* Michx. *Quaking Aspen.* A slender tree reaching a height of about 35 m. and a trunk diameter of 1 m.; bark smooth, light green to white; twigs slender, shiny, reddish-brown; buds conical, brown to black, with 6 or 7 glabrous or ciliate, somewhat gummy scales; leaf-scars conspicuous, lunate; stipule-scars linear, blackish. Dry or moist woods, Labrador to Alaska, south to West Virginia, Illinois, Iowa, New Mexico, Arizona, and Mexico; the most widely distributed tree in North America (Fig. 29).

2. *P. grandidentata* Michx. *Bigtooth Aspen.* A tree 10-25 m. high, 3-6 dm. in diameter; bark smooth, grayish-green; twigs grayish or yellowish-brown, rather thick, glabrous, dull; buds gray, puberulous, with about 6 or 7 visible scales; flowers in catkins, appearing in March. Dry woods and fields, Quebec to Minnesota, south to North Carolina and Missouri (Fig. 30).

3. *P. alba* L. *White Poplar.* A large tree attaining a height of 30 m. and a trunk diameter of 2 m.; bark smooth, light gray; twigs and buds white-tomentose. Introduced from Europe, spreading and naturalized, often too abundant (Fig. 31).

4. *P. deltoides* Marsh. *Eastern Cottonwood.* A large tree, the greatest of the eastern poplars, sometimes 40 m. tall and with a diameter of 2 m.; bark rough on old trees, grayish-green; twigs stout, angular, yellowish-brown, glabrous; lenticels large; terminal buds glabrous, lustrous brown, resinous, with 6 or 7 visible scales, the lateral buds usually smaller, divergent; leaf-scars large, lunate, elevated; stipule-scars dark, conspicuous. River banks and bottomlands, Quebec to Alberta, south to Florida and Texas (Fig. 32).

5. *P. nigra* L. var. *italica* Muenchh. *Lombardy Poplar.* Tree to 30 m. high, the branches closely ascending, forming a nar-

row columnar crown; twigs glabrous, orange, changing to gray; buds slender, glabrous. Introduced from Europe as an ornamental tree, spreading by sprouts; usually staminate. A striking tree because of its formal columnar habit, often planted along avenues (Fig. 33).

6. P. balsamifera L. Balsam Poplar. Tacamahac. (P. tacamahacca Mill.). A tree to 30 m. high, the trunk up to 2 m. in diameter; bark on young stems greenish or reddish-brown, on older trunks becoming gray or grayish-black, divided into scaly or shaggy ridges; buds large, heavily coated with yellow balsam-scented resin; branchlets lustrous, terete. River-banks and rich soil, Labrador to Alaska, south to New York, Michigan, and Colorado.

7. P. gileadensis Rouleau. Balm of Gilead. Tree to 30 m. with stout spreading branches; twigs brown, pubescent; buds large, viscid. Horticultural in origin; only the pistillate tree is known, spreading by sprouts and cuttings (Fig. 34).

8. P. eugenei Saint-Simon. Carolina Poplar. A tree to 30 m. high of pyramidal habit; twigs green, gray or buff; buds small, viscid, tapering from base to apex. Horticultural in origin; only the staminate tree is known, spreading from sprouts and cuttings; formerly much planted as a street tree but now in disfavor because the roots clog sewers (Fig. 35).

9. P. heterophylla L. Swamp Cottonwood. A tree up to 30 m. high; bark furrowed, in narrow plates, somewhat scaly; branchlets whitish-tomentose, becoming glabrate and lustrous; buds 1-1.5 cm. long, canescent-tomentose. Swamps and bottomlands, mostly in the coastal plain, from Louisiana and Florida north to New York, inland about the Great Lakes.

MYRICA L. (Myricaceae)

Aromatic shrubs or small trees, deciduous in cool climates. Twigs rounded or angular, slender, resinous-dotted when young; pith small, somewhat angled, continuous, green. Buds small, solitary, sessile, sub-globose or ovoid, with 2 or 4 exposed scales; end-bud absent. Leaf-scars alternate, half-elliptical or somewhat 3-sided, more or less raised; bundle-traces 3; stipule-scars (if present) small. Fruit globose or ovoid, with a waxy coat or resinous dots.

a. Buds conical or oblong;
fruits covered with resin-drops 1. M. gale

a. Buds subglobose, obtuse; fruits white or drab, encrusted with heavy wax

b. Leaves mostly deciduous; twigs villous when young; fruit 4 mm. in diameter ... 2. M. pensylvanica

b. Leaves evergreen; twigs essentially glabrous

c. Leaves 4-9 cm. long; fruits 2-3 mm. in diameter ... 3. M. cerifera

c. Leaves 1-4 cm. long; fruits 3-4 mm. in diameter ... 4. M. pusilla

1. M. gale L. Sweet Gale. Shrub 0.3-2 m. high, with strongly ascending brown branches; buds conical-ovoid. Swamps, Labrador to Alaska, south to New York, Minnesota and Oregon, and in the Appalachians to North Carolina and Tennessee (Fig. 36).

2. M. pensylvanica Loisel. Bayberry. Candleberry. (M. caroliniensis of authors, not Mill.). Stout stiffly branched shrub 0.3-2 m. high (rarely up to 4.5 m. high, with a trunk 1.2 dm. in diameter); branches mostly whitish-gray or drab, the young ones villous, pilose, or glabrate; buds about 4 mm. long; fruits covered with white wax, 3.5-4.5 mm. in diameter. Sterile soil, mostly in the coastal plain, North Carolina to Newfoundland, inland about Lake Erie (Fig. 37).

3. M. cerifera L. Wax-Myrtle. Shrub or tree up to 12 m. high, with a trunk diameter up to 2 dm.; young branches waxy, glabrous or sparsely pilose; leaves evergreen, narrowly oblanceolate, 4-9 cm. long and 0.5-2 cm. broad, yellow-green, coriaceous, heavily coated with waxy granules; buds small, about 1 mm. long, glandular-dotted; fruit 2-3 mm. in diameter. Thickets and swamps, Florida to Texas, north to New Jersey and Arkansas; mostly on the coastal plain (Fig. 38).

4. M. pusilla Raf. Dwarf Wax-Myrtle. Low colonial stoloniferous shrub 0.2-2 m. high; branchlets waxy, glabrous or nearly so; leaves evergreen, coriaceous, oblanceolate to obovate and obtuse, 1.5-4 cm. long; fruits 3-4 mm. in diameter. Pine barrens and woods, Florida to Texas, north to Dèlaware and Arkansas; mostly on the coastal plain (Fig. 39).

Fig. 35. Populus eugenei

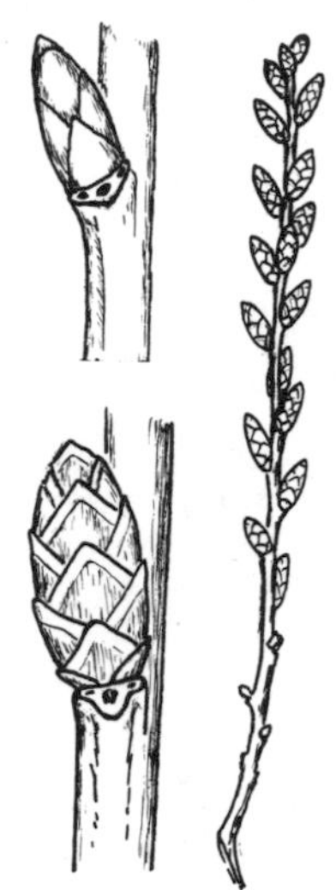

Fig. 36. Myrica gale

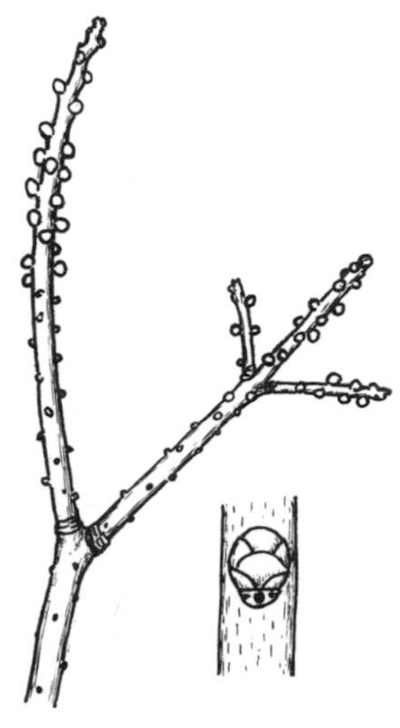

Fig. 37. Myrica pensylvanica

Fig. 38. Myrica cerifera

Fig. 39. Myrica pusilla

COMPTONIA L'Her. (Myricaceae)

Low pubescent deciduous shrub, with fragrant leaves, fruit and twigs. Fruit a conical hard smooth nut, surrounded by 8 long linear bracts persisting as a sort of bur.

1. C. peregrina (L.) Coult. Sweet Fern. (Myrica asplenifolia L.). A shrub 3-6 dm. high, the branches erect or spreading; twigs slender, resinous-dotted when young; pith small, solitary, sessile, ovoid, with about four exposed scales; end-bud absent; leaf-scars alternate, 3-sided, somewhat raised; bundle-traces 3; stipule-scars small. The leaves are sweet-scented and fern-like in appearance; they are deciduous in winter but a few withered ones may generally be found. Open woodlands and barrens, Nova Scotia to Manitoba, south to Minnesota, Indiana and in the mountains to Georgia and Tennessee (Fig. 40).

JUGLANS L. (Juglandaceae)

Trees with spreading branches, superposed buds, fragrant bark and brown chambered pith. Leaf-scars alternate, shield-shaped or 3-lobed, large; bundle-traces in 3 compound groups; stipule-scars none. The fruit, present in winter, may be classed as a tryma (see p. 11).

a. Terminal bud oblong, elongated; leaf-scar with a downy line across the top; pith dark brown. 1. J. cinerea
a. Terminal bud ovoid or subglobose; leaf-scar notched at top, without a downy line; pith light brown. 2. J. nigra

1. J. cinerea L. Butternut. White Walnut. A tree 16-30 m. high, 6-9 dm. in diameter, with an open crown of spreading branches; bark gray, on old trunks divided by fissures into lighter flat-topped ridges; leaf-scars not notched; pith dark brown. Rich woods, New Brunswick to North Dakota, south to Georgia and Arkansas (Fig. 41).

2. J. nigra L. Black Walnut. A handsome tree 20-35 m. high, 6-18 dm. in diameter; crown round, open; bark dark brown with deep furrows; leaf-scars notched at the top; pith light brown. Rich woods, Massachusetts to Minnesota, south to Georgia and Arkansas (Fig. 42).

CARYA Nutt. (Juglandaceae)

Deciduous trees with close or shaggy bark, terete twigs and continuous pith. Buds large, sometimes stalked, often superposed. Leaf-scars alternate, shield-shaped or 3-lobed, large; bundle-traces numerous, mostly in 3 more or less definite groups. The

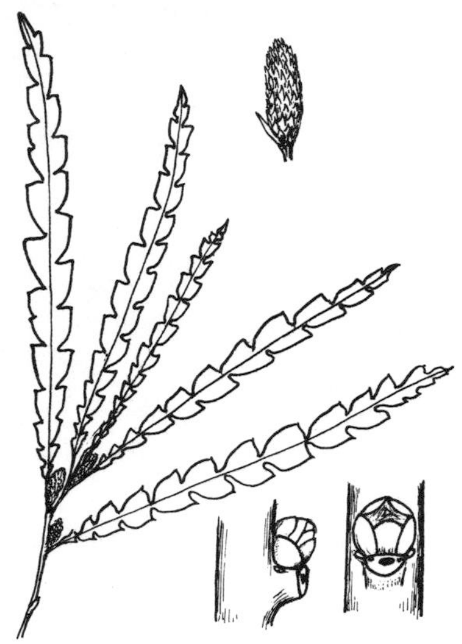

Fig. 40. Comptonia peregrina

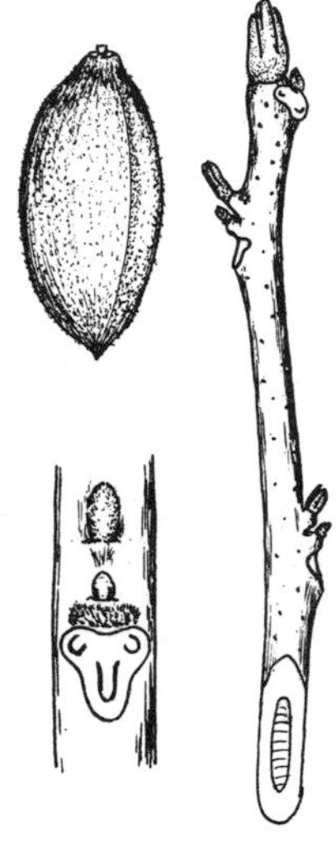

Fig. 41. Juglans cinerea

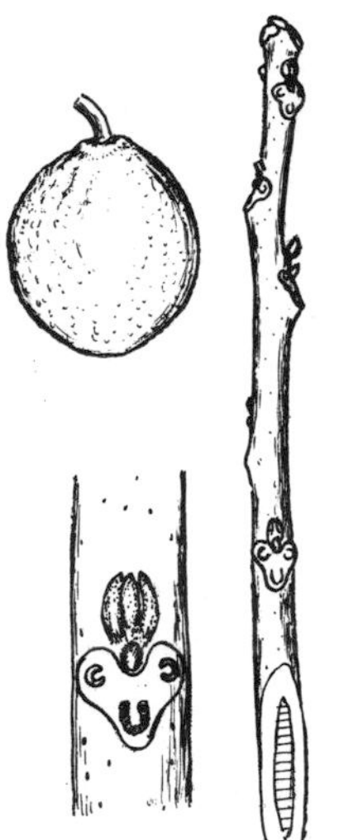

Fig. 42. Juglans nigra

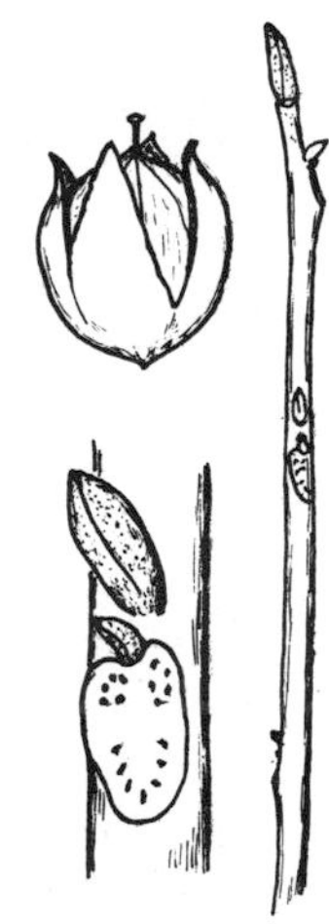

Fig. 43. Carya cordiformis

Fig. 44. Carya ovata

fruit, present in winter, either on the tree, or mostly scattered on the ground, may be classed as a kind of dry drupe, or <u>tryma</u>. <u>Hicoria</u> Raf.

a. Bud-scales valvate, in pairs

  b. Buds covered with yellow-brown hairs; kernel sweet — 1. C. illinoensis

  b. Buds sulfur-yellow, scurfy; kernel bitter — 2. C. cordiformis

a. Bud-scales imbricate, usually more than 2 exposed

  b. Terminal bud large, usually over 10 mm. long, mostly hairy; fruit with a thick husk

    c. Outer bud-scales persistent, the bud shaggy in appearance

      d. Twigs dark reddish-brown; fruit 3-6 cm. long — 3. C. ovata

      d. Twigs orange-brown or buff-colored; fruit 4-6.5 cm. long — 4. C. laciniosa

    c. Outer bud-scales falling early, the bud neat in appearance — 5. C. tomentosa

  b. Terminal bud small, usually less than 10 mm. long; mostly glabrous or scaly; fruit with a thin husk

    c. Terminal bud essentially glabrous (hairy if the outer bud-scales have fallen)

      d. Fruit obovoid, the husk usually splitting only part way down (or tardily splitting) — 6. C. glabra

      d. Fruit usually ellipsoid or subglobose, the husk splitting promptly to the base — 7. C. ovalis

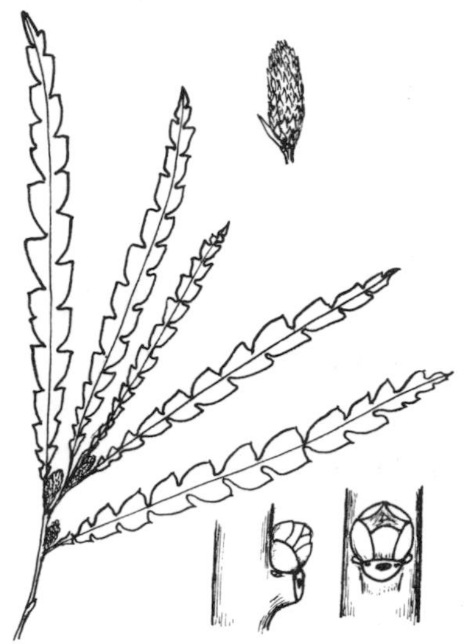

Fig. 40. Comptonia peregrina

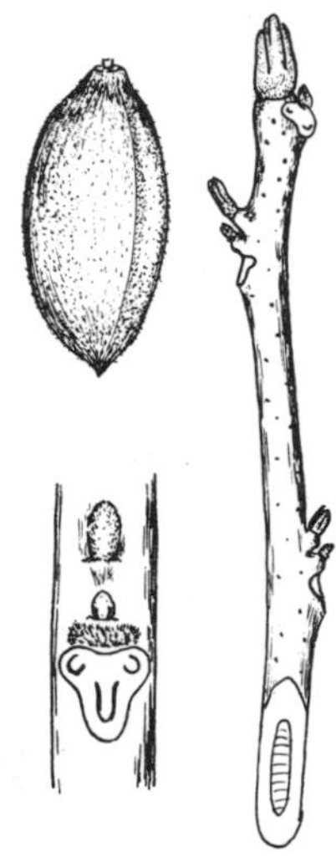

Fig. 41. Juglans cinerea

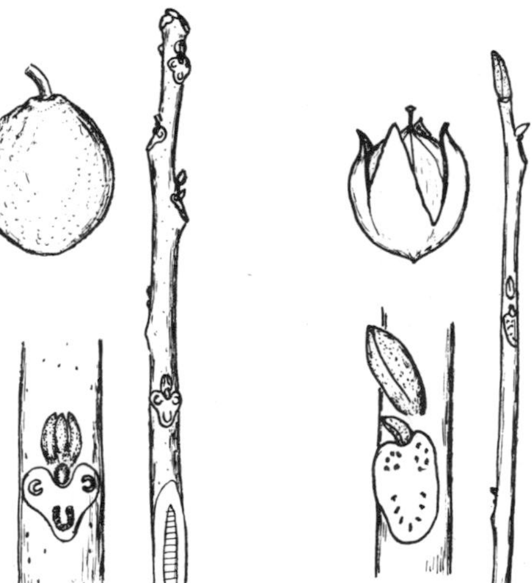

Fig. 42. Juglans nigra

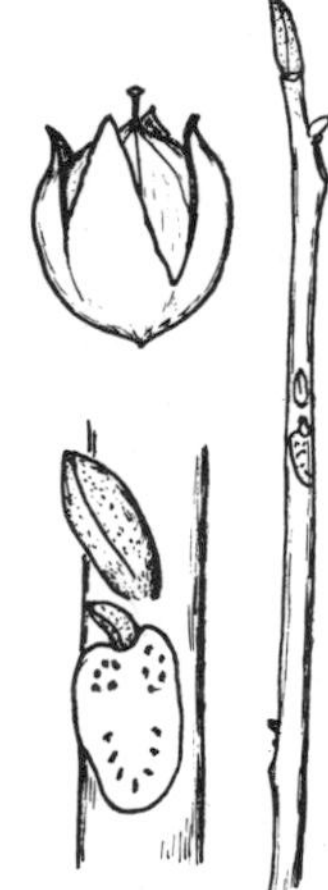

Fig. 43. Carya cordiformis

Fig. 44. Carya ovata

c. Terminal bud covered with silvery scales 8. C. pallida

1. C. illinoensis (Wang.) K. Koch. Pecan. (C. pecan (Marsh.) Engl. and Graebn.). Tree up to 50 m. high, and with a trunk diameter of 6-12 dm., buttressed at the base; bark very thick, furrowed, ridged; buds flattened, with paired valvate narrow scales covered with jointed hairs; fruit elongate, the ripe husk splitting to below the middle; nut reddish-brown, elongate, subcylindric, ellipsoid or ovoid; shell thin; kernel sweet. Bottomlands, Indiana to Iowa, south to Alabama and Texas.

2. C. cordiformis (Wang.) K. Koch. Bitternut Hickory. A slender tree 15-50 m. high, the trunk 3-7.5 dm. in diameter; bark close and rough; bud-scales valvate in pairs, bright yellow; husk of fruit thin, splitting slowly into 4 valves; nut globular, thin-walled; seed extremely bitter. Woods, Florida to Texas, north to New Hampshire, Minnesota, and Nebraska (Fig. 43).

3. C. ovata (Mill.) K. Koch. Shagbark Hickory. A handsome tree 20-28 m. high, the trunk 3-6 dm. in diameter; bark shaggy, exfoliating in rough strips; twigs gray or brown, puberulent or glabrate; bud scales several, imbricated; fruit-husk thick, splitting quickly into four valves when mature; nut flattish-globular, thin-walled; seed sweet. Rich woods and slopes, Florida to Texas, north to Maine, Minnesota, and Nebraska (Fig. 44).

4. C. laciniosa (Michx. f.) Loud. Shellbark Hickory. A tree 20-40 m. high, the trunk 3-6 dm. in diameter; bark separating into long narrow straight plates; twigs buff or orange; fruit-husk thick, splitting quickly into four valves when mature; nut large, 3-5 cm. long, angular, thick-walled; seed sweet. Bottomlands, New York to Iowa and Nebraska, south to Alabama and Louisiana (Fig. 45).

5. C. tomentosa Nutt. Mockernut. White Hickory. (C. alba K. Koch). A tree 20-40 m. high, the trunk 3-7.5 dm. in diameter; bark close, rough, but not shaggy or exfoliating; twigs tomentose-pubescent; outer bud-scales falling early, giving the bud a clean neat silvery appearance; fruit-husk thick and hard, splitting to the base; nut globular, quite thick-shelled (hence the common name, mockernut); seed sweet. Woodlands, Florida to Texas, north to Vermont, Michigan, and Nebraska (Fig. 46).

6. C. glabra (Mill.) Sweet. Pignut Hickory. A tree 20-30 m. high, the trunk 3-10 dm. in diameter; bark close, rough; twigs glabrous or nearly so; terminal bud small, less than 10 mm. long;

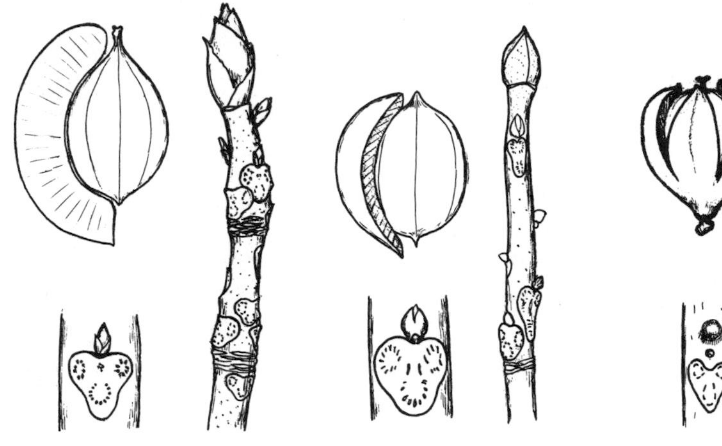

Fig. 45. Carya laciniosa

Fig. 46. Carya tomentosa

Fig. 47. Carya glabra

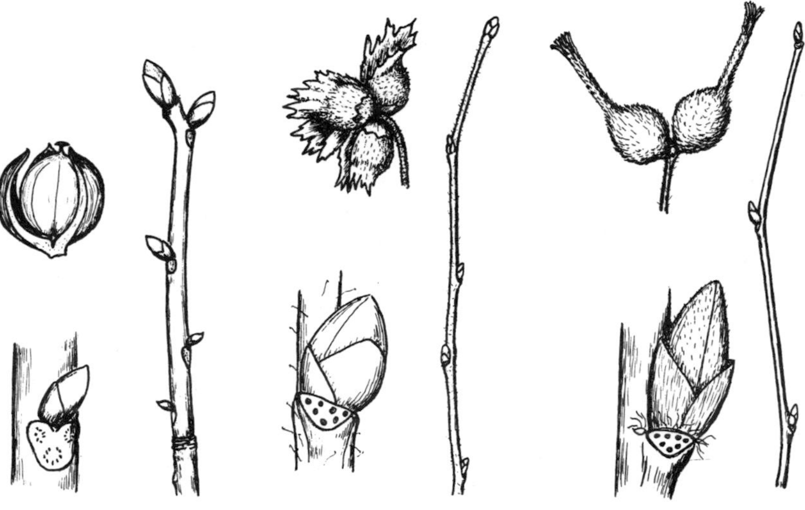

Fig. 48. Carya ovalis

Fig. 49. Corylus americana

Fig. 50. Corylus cornuta

fruit obovoid, the husk thin, splitting only part way to the base;nut rather thick-shelled; seed rather bitter. Dry woods, Vermont to Illinois, south to Florida and Louisiana (Fig. 47).

7. C. ovalis (Wang.) Sarg. Oval Pignut Hickory. Red Hickory. A tree 20-40 m. in height, 3-10 dm. in diameter; bark closely and deeply furrowed; twigs red-brown, glabrous; terminal bud small, less than 10 mm. long; fruit oval in vertical section, the husk thin, splitting freely to the base; nut 4-ribbed above the center; kernel rather sweet. Rich soil, Massachusetts to Michigan and Iowa, south to Georgia and Arkansas (Fig. 48).

8. C. pallida (Ashe) Engl. and Graebn. Pale Hickory. Sand Hickory. A tree usually 10-15 m. high, with trunk diameter of 3-6 dm.; bark very pale to dark gray, becoming deeply furrowed; buds silvery, the terminal about 6 mm. long; fruit yellow-scaly, 1.5-4 cm. long, the husk thin, tardily splitting to base; nut thin-shelled; kernel sweet. Dry sandy soil, mostly on or near the coastal plain, Louisiana to Florida, north to Tennessee and New Jersey.

CORYLUS L. (Betulaceae)

Deciduous shrubs. Twigs moderate, terete, zigzag; pith continuous, 3-sided, pale or brown. Buds solitary, sessile, ovoid, obtuse, with about 4-6 scales. Leaf-scars alternate, 2-ranked, more or less triangular, small; bundle-traces 3 or multiplied; stipule-scars elongated.

a. Bud-scales persistent, the outer short; fruit-involucre broad — 1. C. americana
a. Outer bud-scales elongated, often deciduous; fruit-involucre prolonged into a slender beak — 2. C. cornuta

1. C. americana Walt. American Hazelnut. A shrub 1-3 m. high, young shoots russet-brown, densely bristly with pinkish hairs becoming darker, mostly falling before spring; buds 4 mm. long, pubescent; involucre composed of two broad bracts; nut light brown, 1.1-5 cm. long. Thickets, Maine to Saskatchewan, south to Georgia and Oklahoma (Fig. 49).

2. C. cornuta Marsh. Beaked Hazelnut. (C. rostrata Ait.). Shrub 1-3 m. high; twigs pubescent; involucre composed of united bracts prolonged above the nut into a narrow tubular beak. Thickets, Newfoundland to British Columbia, south to Georgia, Kansas, and Colorado (Fig. 50).

OSTRYA Scop. (Betulaceae)

Deciduous trees with scaly bark. Twigs slender, zigzag, terete; pith small, pale green, round or somewhat angled, continuous. Buds usually solitary, sessile, ovoid, oblique, with about 6 spirally arranged scales, striped with faint longitudinal grooves; terminal bud absent. Leaf-scars alternate, crescent-shaped or elliptical, small; bundle-traces 3; stipule-scars unequal.

1. O. virginiana (Mill.) K. Koch. Hop Hornbeam. Ironwood. A slender tree up to 20 m. high, the trunk 6 dm. or less in diameter, with very hard wood and brownish furrowed bark; twigs pubescent; staminate catkins sessile at ends of twigs of the preceding season; pistillate catkins small, terminal, erect, the flowers subtended by a tubular bract which in fruit becomes a closed bladdery bag 12-15 mm. long; ripe pistillate catkins resemble hops, whence the common name. Rich woods, Nova Scotia to Manitoba and South Dakota, south to Florida and Texas (Fig. 51).

CARPINUS L. (Betulaceae)

Deciduous trees with smooth dark bluish-gray bark, the trunk with irregular ridges extending up and down ("muscle-like"). Twigs slender, zigzag, terete; pith small, round or 5-sided, continuous, pale. Buds usually solitary, ovoid, sessile, oblique, with about twelve 4-ranked scales. Leaf-scars alternate, small, crescent-shaped; bundle-traces 3; stipule-scars unequal.

1. C. caroliniana Walt. American Hornbeam. Blue-Beech. Water-Beech. Muscle-Tree. A small tree to 15 m. high, trunk short, often leaning, furrowed and ridged; bark dark bluish-gray, smooth, thin, tight; twigs glabrous or nearly so; wood very hard; buds brown, somewhat silky, less than 5 mm. long; staminate catkins entirely enclosed in buds, not visible in winter; fruit subtended by a flat persistent bract which becomes much enlarged (foliaceous) and lobed or incised; nutlet 4 mm. long. Rich woods, New England to Ontario and Minnesota, south to Florida and Texas (Fig. 52).

BETULA L. (Betulaceae)

Aromatic deciduous trees or shrubs. Twigs slender, terete; lenticels elongated horizontally (in this respect resembling cherry); pith minute, flattened, continuous, green. Buds moderate, solitary, fusiform-ovoid, sessile, with 2 or 3 visible scales. Leaf-scars alternate, oval, triangular or crescent-shaped; bundle-traces 3. Staminate catkins naked through winter, often 2 or 3 in a cluster; pistillate catkins ovoid or cylindrical.

a. Twigs with a sweet wintergreen fragrance

b. Bark on old trunks black; buds sharply

pointed, divergent, mostly glabrous; twigs brown to black, quite aromatic 1. B. lenta

b. Bark on old trunks bronze, scaly; buds appressed, at least along the lower half, often hairy; twigs greenish-brown, somewhat aromatic 2. B. lutea

a. Twigs without a wintergreen fragrance

b. Bark on old trunks grayish-white, usually close; twigs gray, with prominent warty lenticels; buds short, tapering both ways from the middle 4. B. populifolia

b. Bark on old trunks exfoliating in thin strips; twigs brown to black; buds tapering from base to apex

c. Twigs and buds somewhat hairy; twigs reddish-brown; freshly exposed bark on old trunks salmon-pink or greenish-brown, the older dark. 3. B. nigra

c. Twigs and buds mostly glabrous; twigs nearly black; bark on old trunks chalky-white 5. B. papyrifera

1. B. lenta L. Sweet Birch. Black Birch. Cherry Birch. A tree 15-20 m. high, 6-12 dm. in diameter; bark dark-brown, close, smooth, becoming furrowed, not separating in layers, the young bark having a very sweet wintergreen taste; lenticels prominent; twigs glabrous, slender and pliable, red-brown; fruiting catkins short-cylindric, 1.5-3.5 cm. long, the scales firm and smooth; nut oblong, with narrow wings. Rich woods, Maine to Ontario, south to Georgia and Tennessee (Fig. 53).

2. B. lutea Michx. f. Yellow Birch. A tree 20-30 m. high, and 6-12 dm. in diameter; bark of trunk yellowish or silvery-gray, detaching in thin filmy layers; twigs gray-brown, somewhat aromatic; fruiting catkins oblong, 2-4 cm. long, the scales pubescent; nut oblong, with narrow wings. Rich cool woods, Nova Scotia to Ontario, south to North Carolina and Iowa (Fig. 54).

3. B. nigra L. River Birch. Red Birch. A tree 15-30 m. high, 3-9 dm. in diameter; bark reddish-brown, deeply furrowed and broken into thin irregular scales, exposing the orange-red close bark underneath; twigs reddish; buds hairy, the lower scales elongated; fruiting catkins thick-cylindric, 2.5-3.5 cm. long, the

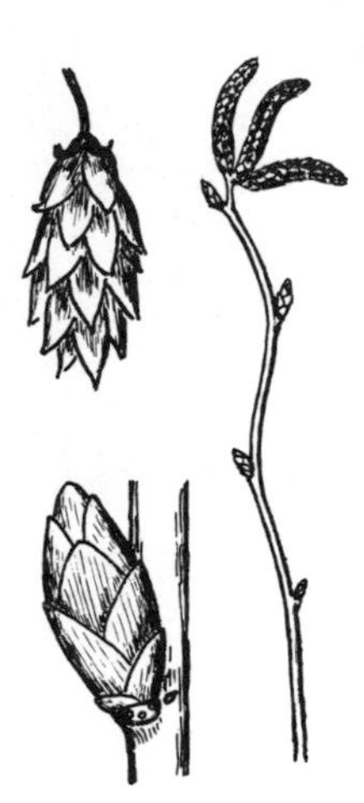

Fig. 51. Ostrya virginiana

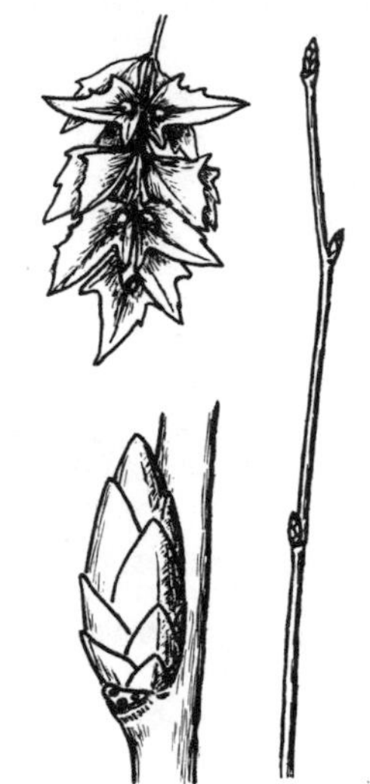

Fig. 52. Carpinus caroliniana

Fig. 53. Betula lenta

Fig. 54. Betula lutea

Fig. 55. Betula nigra

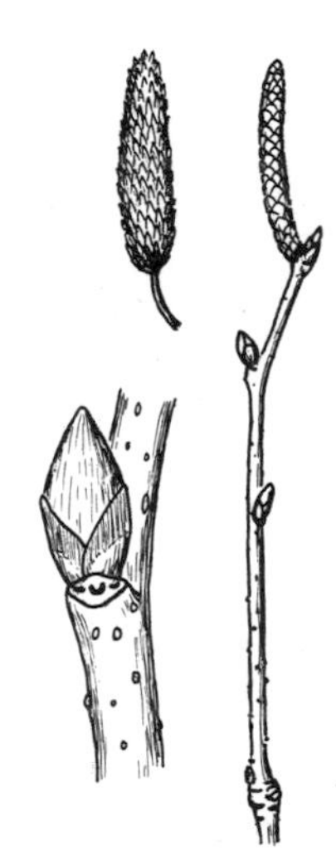

Fig. 56. Betula populifolia

scales tomentose; nut broadly ovate, winged, pubescent. Bottomlands, Florida to Texas, north to New Hampshire, Michigan, Minnesota, and Kansas (Fig. 55).

4. B. populifolia Marsh. Gray Birch. Bushy tree up to 20 m. high; bark close, not exfoliating, chalky-white with dark elongate markings; twigs slender, glabrous, wiry, often resinous-warty; buds ovoid, glabrous, about 6 mm. long; fruiting catkins 1.8 cm. long. Sterile soil, Prince Edward Island to Quebec, south to Delaware and Indiana (Fig. 56).

5. B. papyrifera Marsh. Paper Birch. Canoe Birch. A tree up to 25 m. high, with a trunk diameter of 1 m.; bark lustrous, white or bronze, peeling in thin layers; buds glabrous; fruiting catkins 1.5-4.5 cm. long, spreading or drooping on slender peduncles, the scales ciliate-margined. Woods, Labrador to Alaska, south in the Appalachians to West Virginia and North Carolina, in the western mountains to Montana and Washington (Fig. 57).

ALNUS B. Ehrh. (Betulaceae)

Deciduous shrubs with gray bark. Twigs somewhat 3-sided; pith small, 3-sided, continuous. Buds sessile or stalked, rather large, solitary, with 3 scales. Leaf-scars alternate, half-round, raised; bundle-traces 3 or compound. Both kinds of flowers in catkins, the staminate for next season usually conspicuous in winter, along with the persistent cone-like fruiting catkins of the preceding season and the small undeveloped pistillate catkins of the next season.

a. Buds sessile, with 3-6 imbricated scales — 1. A. crispa
a. Buds stalked, with 2 or 3 scales
  b. Stems with linear transverse lenticels; fruiting catkins bent downwards — 2. A. rugosa
  b. Lenticels shorter, fewer, darker; fruiting catkins erect — 3. A. serrulata

1. A. crispa (Ait.) Pursh. Green Alder. Ascending and bushy shrub up to 3 m. high; young branches glabrous or sparsely pubescent; buds sessile. Labrâdor to Alaska, south to North Carolina, Michigan, Minnesota, and Alberta (Fig. 58).

2. A. rugosa (DuRoi) Spreng. Speckled Alder. Hoary Alder. (A. incana of authors, not (L.) Moench.). A shrub or small tree 2-8 m. high, the twigs glabrous in winter; trunk marked with

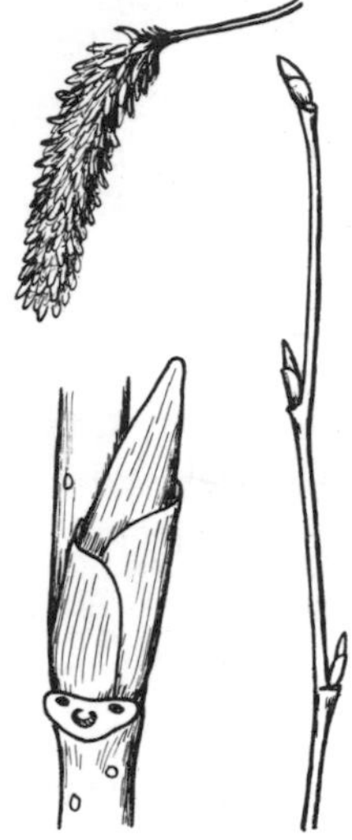
Fig. 57. Betula papyrifera

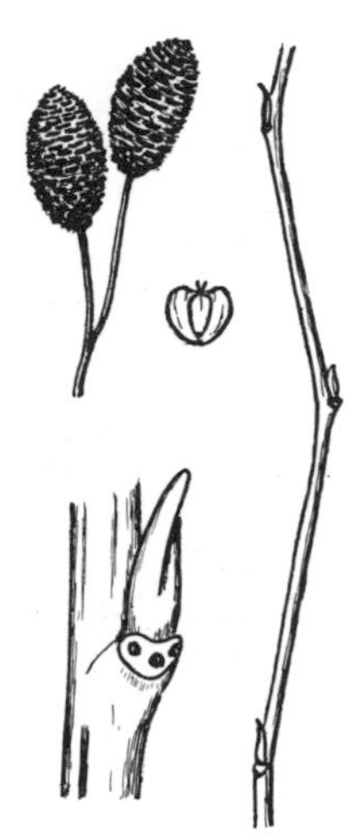
Fig. 58. Alnus crispa

Fig. 59. Alnus rugosa

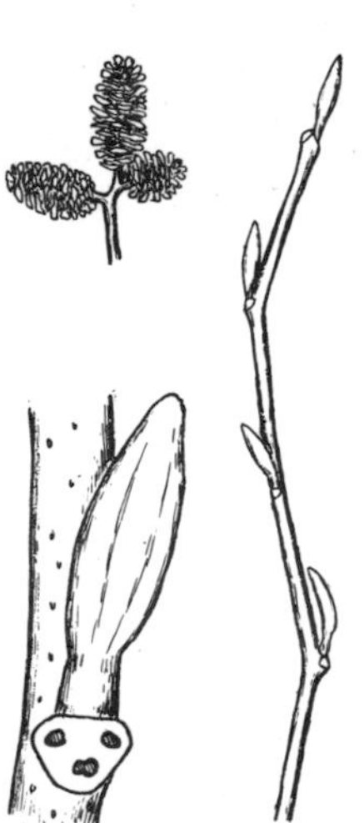
Fig. 60. Alnus serrulata

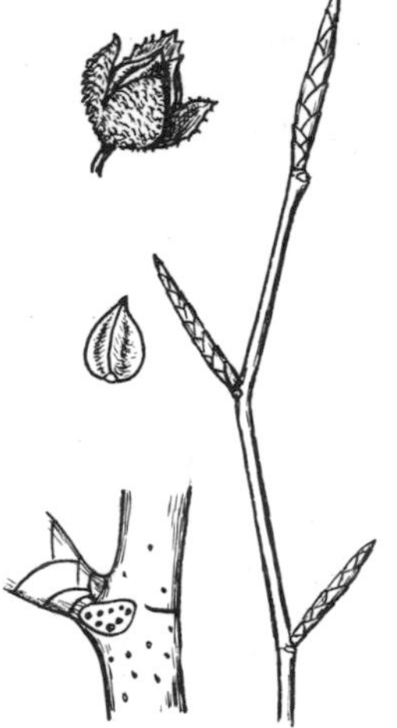
Fig. 61. Fagus grandifolia

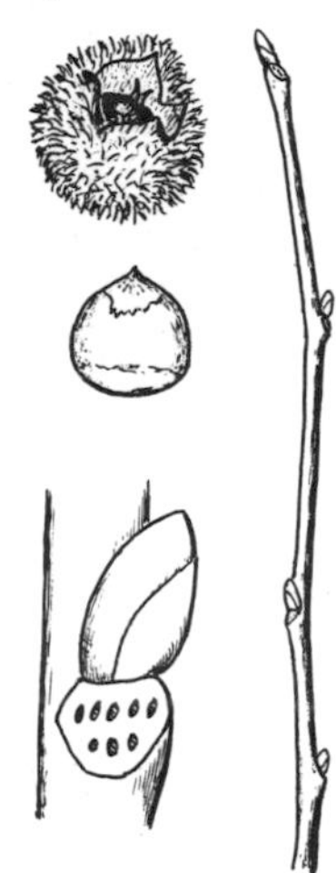
Fig. 62. Castanea dentata

whitish linear lenticels up to 7 mm. long (hence the name, speckled alder); fruiting catkins ovoid, 12 mm. long, bent downwards; nut orbicular. The flowers begin to appear in March. Low grounds, Labrador to Saskatchewan, south to West Virginia, Indiana, and Iowa; the common species northwards (Fig. 59).

3. A. serrulata (Ait.) Willd. Smooth Alder. Brookside Alder. (A. rugosa of authors, not Spreng.). A shrub or small tree 2-7 m. high, the bark smooth; fruiting catkins erect, ovoid, 12-18 mm. long; nut ovate. The flowers begin to appear in February and March. Stream-margins, Florida to Louisiana, north to Nova Scotia, Ohio, Missouri, and Oklahoma; the common species southwards (Fig. 60).

FAGUS L. (Fagaceae)

Deciduous trees with smooth gray bark, even on large trunks. Twigs slender, terete, zigzag; pith small, round, continuous. Buds usually solitary, sessile or short-stalked, elongate-fusiform, divergent, oblique over the leaf-scar, with 10 or more spirally arranged scales. Leaf-scars small, alternate, half-round; bundle-traces 3 or multiplied; stipule-scars linear, nearly meeting around the twig. Fruit a nut, usually 2 together in an urn-shaped prickly involucre which splits into 4 valves.

1. F. grandifolia Ehrh. Beech. A tree 15-36 m. high, with a trunk diameter of 6-9 dm.; bark smooth, close, light gray, often with included woody nodules; twigs glabrous; buds red-brown. Rich woods, Prince Edward Island to Ontario and Illinois, south to Florida and Texas (Fig. 61).

CASTANEA Mill. (Fagaceae)

Deciduous shrubs or trees with fissured bark. Twigs moderate, somewhat grooved; pith moderate, continuous, angled. Buds solitary, ovoid, oblique, sessile, with 2 or 3 visible scales, the terminal bud sometimes absent. Leaf-scars alternate, half-round; bundle-traces 3 or much divided; stipule-scars unequal. Nuts usually 3 together in a prickly involucre ("bur").

a. Twigs and buds glabrous, or with scattered hairs — 1. C. dentata
a. Twigs and buds grayish-tomentose — 2. C. pumila

1. C. dentata (Marsh.) Borkh. Chestnut. A tree 20-30 m. high, 1-2 m. thick, with brown bark in longitudinal plates; twigs chestnut-brown; "bur" 4-10 cm. in diameter, enclosing 1-5 nuts; nuts 1-2.5 cm. wide, quite sweet. Dry mostly acid soil, Maine and Minnesota, south to Georgia and Mississippi; now mostly de-

stroyed by chestnut blight (Fig. 62).

2. C. pumila (L.) Mill. Chinquapin. A spreading shrub or small tree 6–10 m. high, 3–6 dm. in diameter; bark lightly furrowed, with flat ridges broken into light brown loose plates; "bur" 4 cm. in diameter or less; nut ovoid, not flattened, quite sweet, but scarcely half as large as the chestnut. Dry woods and thickets, Florida to Texas, north to Massachusetts, Tennessee, and Arkansas (Fig. 63).

QUERCUS L. (Fagaceae)

Mostly deciduous trees or shrubs, the dried leaves often somewhat persistent through the winter (marcescent). Twigs moderate or slender, grooved; pith moderate, continuous, star-shaped in cross section. Buds solitary, sessile, clustered toward the tips of the branches, with numerous scales. Leaf-scars alternate, half-round; bundle-traces numerous, scattered; stipule-scars small. Fruit an acorn (a nut surrounded at its base by a cup-like involucre). The species hybridize freely and are sometimes difficult to distinguish. In addition to bud and twig characters, the key makes use of leaves and fruits, some of which are often present in winter.

a. Leaves deciduous

b. Largest terminal buds mostly 6–9 mm. long, usually acute

c. Buds distinctly angled in cross section

d. Buds mostly glabrous, dull straw-colored ... 13. Q. shumardii

d. Buds pubescent, dark red or gray

e. Buds whitish-pubescent near the apex, sometimes obtuse ... 12. Q. coccinea

e. Buds gray—or rusty-tomentose, long (often 10 mm. long), acute

f. Buds gray-tomentose; twigs often shiny ... 14. Q. velutina

f. Buds rusty-tomentose; twigs usually dull, often scurfy-pubescent ... 17. Q. marilandica

c. Buds circular or only slightly angled in cross section.

d. Buds and twigs orange-brown; buds about 6 mm. long, slender, acute — 8. Q. prinus

d. Twigs reddish-brown; buds reddish-brown or nearly black; plump, obtuse or acute

e. Buds reddish-brown; scales near the apex, silky, or puberulous, mostly on the margins

f. Leaves oblong, lobed — 10. Q. rubra

f. Leaves obovate, crenate-dentate — 6. Q. michauxii

e. Buds reddish-brown to nearly black; scales near the apex whitish-pubescent — 12. Q. coccinea

b. Largest terminal buds mostly less than 6 mm. long, acute or obtuse

c. Buds mostly acute

d. Twigs gray-tomentose, becoming glabrate

e. Straggling shrub, mostly in the mountains — 16. Q. ilicifolia

e. Large timber tree, not in the mountains — 15. Q. falcata

d. Twigs mostly glabrous

e. Buds and twigs brown or orange-brown — 7. Q. muehlenbergii

e. Buds and twigs red or reddish-brown

f. Bud-scales glabrous; leaves lobed — 11. Q. palustris

f. Bud-scales pubescent; leaves entire — 18. Q. imbricaria

c. Buds obtuse, often nearly globose

d. Twigs shiny, red or reddish-brown

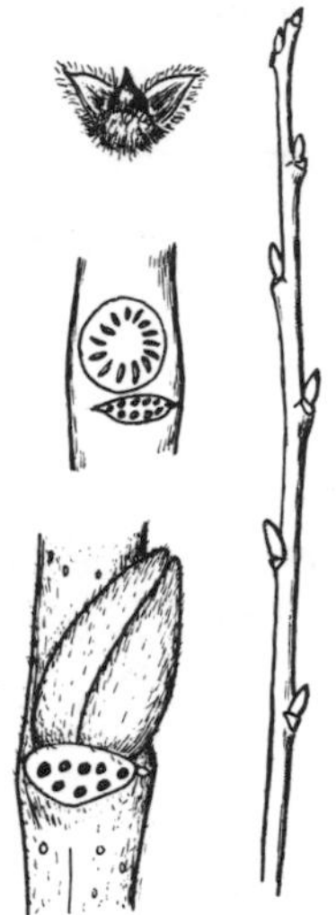

Fig. 63. Castanea pumila

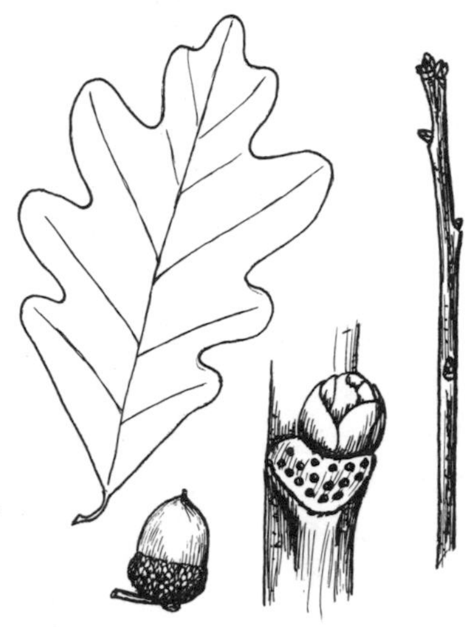

Fig. 64. Quercus alba

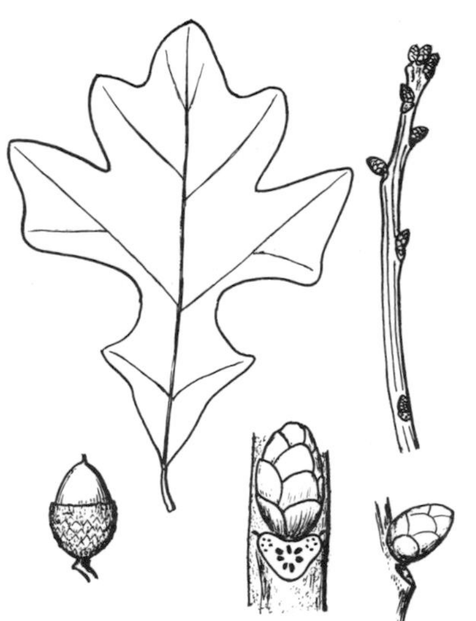

Fig. 65. Quercus stellata

Fig. 66. Quercus lyrata

e. Buds globose or nearly so; bud-scales glabrous; twigs reddish — 1. Q. alba
e. Buds obtuse to somewhat acute, often angled, whitish-pubescent near the apex; twigs reddish-brown — 12. Q. coccinea

d. Twigs dull, yellowish-brown

e. Buds and twigs glabrous — 5. Q. bicolor
e. Buds and twigs more or less tomentose

f. Buds about 1.5 mm. long; twigs slender — 16. Q. ilicifolia
f. Buds about 3 mm. long; twigs moderate

g. Acorn 2.5-3.5 cm. long

h. Cup of the acorn prominently fringed on the margin — 4. Q. macrocarpa
h. Cup of the acorn not fringed — 3. Q. lyrata

g. Acorn about 12 mm. long — 2. Q. stellata

a. Leaves evergreen — 9. Q. virginiana

1. Q. alba L. White Oak. A tree 25-35 m. high, 1-2 m. in diameter; bark gray, rough; twigs gray or purple, often glaucous; buds deep-brown, medium-sized, subglobose or ellipsoid, nearly or quite glabrous, about 5 mm. long; stipules glaucous; acorn ovoid or ellipsoid, 2-3 cm. long, the cup hemispherical, warty, much shorter than the nut. Rich woods, Maine to Minnesota, south to Florida and Texas (Fig. 64).

2. Q. stellata Wang. Post Oak. A tree 15-25 m. high, 6-9 dm. in diameter, with rough gray bark; buds ovoid or conical-ovoid, dull, silky, 3 mm. long; twigs yellow-scurfy; acorn 1-2 cm. long. Sterile soil, Massachusetts to Illinois and Kansas, south to Florida and Texas (Fig. 65).

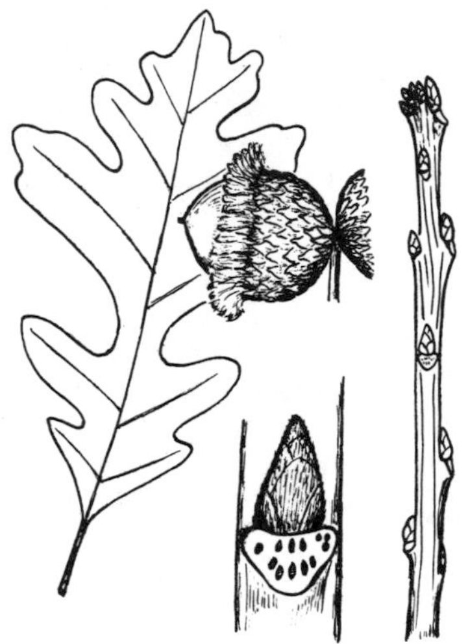

Fig. 67. Quercus macrocarpa

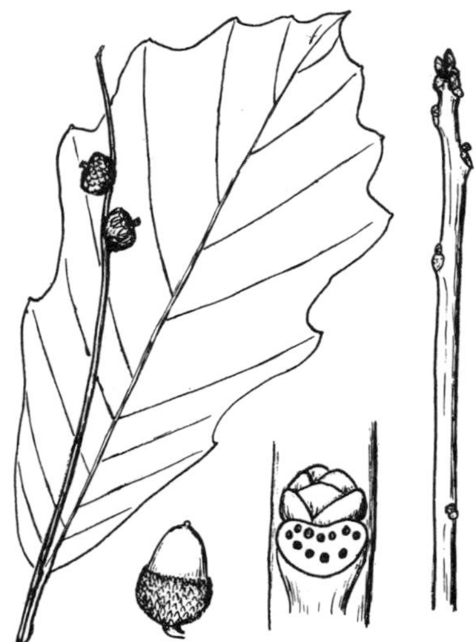

Fig. 68. Quercus bicolor

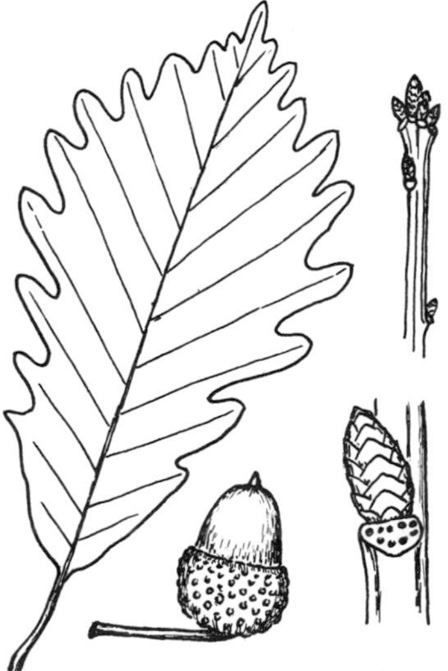

Fig. 69. Quercus michauxii

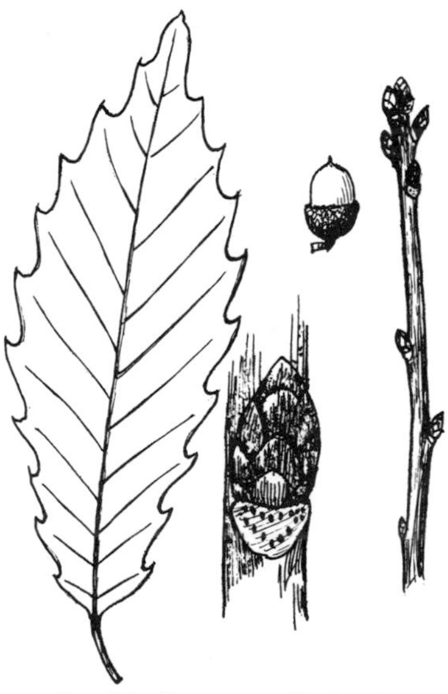

Fig. 70. Quercus muehlenbergii

3. Q. lyrata Walt. Overcup Oak. A tree up to 35 m. high, with a trunk diameter of 9-12 dm.; bark gray; twigs buff, pubescent, becoming glabrate; buds 3 mm. long, ovoid-globose, the scales light chestnut-brown, somewhat tomentose; acorn 1.2-2.5 cm. long, the cup subglobose, pubescent, nearly covering the depressed nut. Bottomlands, mostly in the coastal plain, Texas and Florida to New Jersey, up the Mississippi Valley to Indiana and Missouri (Fig. 66).

4. Q. macrocarpa Michx. Bur Oak. A tree 15-40 m. high, 1-2 m. in diameter, with gray flaky bark; buds ovoid or conical-ovoid; twigs and buds gray-pubescent; acorn 1.5-3.5 cm. high, the nut half-covered or rarely nearly covered by the fringed bur-like cup. Bottomlands, New Brunswick to Manitoba, south to North Carolina, Tennessee, and Texas; not in the coastal plain (Fig. 67).

5. Q. bicolor Willd. Swamp White Oak. A tree 15-25 m. high, 6-9 dm. in diameter, with flaky gray bark, exfoliating from the branches; twigs straw-brown; buds 2-3 mm. long, subglobose or ellipsoid, nearly or quite glabrous; stipule-scars lacking or inconspicuous; acorn 2-3 cm. long, the cup one-third to one-half as long as the nut. Bottomlands, Maine to Minnesota and Nebraska, south to Georgia, Kentucky, and Oklahoma (Fig. 68).

6. Q. michauxii Nutt. Basket Oak. Cow Oak. Swamp Chestnut Oak. (Q. prinus of authors, not L.). Tree 20-30 m. high, with a trunk diameter of 6-9 dm.; bark silvery-whitish; twigs olive or brown; buds 6 mm. long, the scales red, puberulous; acorn ovoid-cylindric, about 3 cm. long. Bottomlands and swamps, mostly in the coastal plain, Texas to Florida and New Jersey, north in the Mississippi Valley to Indiana and Missouri (Fig. 69).

7. Q. muehlenbergii Engelm. Yellow Oak. Muhlenberg Oak. Chinquapin Oak. A tree 15-25 m. high, 6-9 dm. in diameter, with close thin gray bark, flaky when old; twigs orange-brown; buds 3 mm. long, ovoid or conical-ovoid, brown-puberulent or glabrous, the scales often pale-margined; acorn globose or obovoid, 1.2-2 cm. long, sessile or nearly so, the cup thin, enclosing half the nut. Dry limestone slopes, Vermont and Minnesota, south to Florida and Texas (Fig. 70).

8. Q. prinus L. Chestnut Oak. Rock Oak (Q. montana Willd.) A tree 20-30 m. high, 1-2 m. in diameter, with thick and deeply grooved bark; twigs brown; buds distinctly conical, 5-6 mm. long, deep brown, dull, outer scales pale-margined; acorn 2.5-3.5 cm. long; ovoid, the cup thick. Dry woods, mostly on sandstone, Maine to Indiana, south to Georgia and Mississippi; not in the coastal

Fig. 71. Quercus prinus

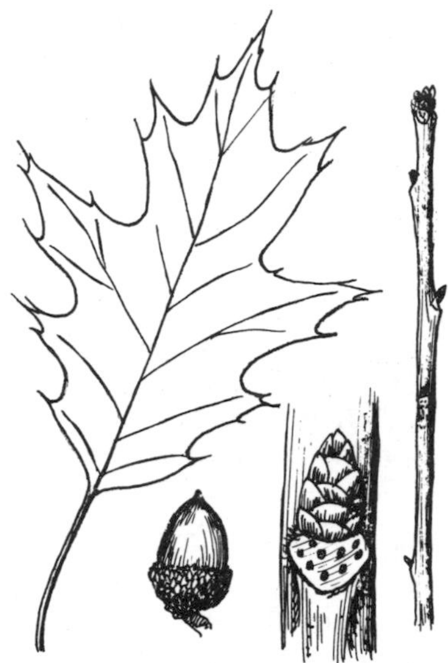
Fig. 72. Quercus rubra

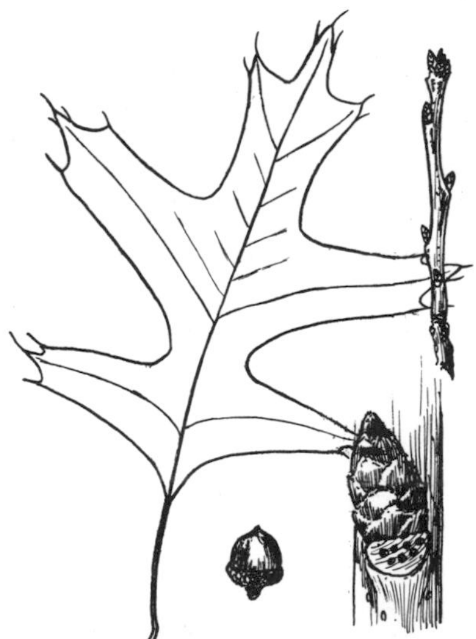
Fig. 73. Quercus palustris

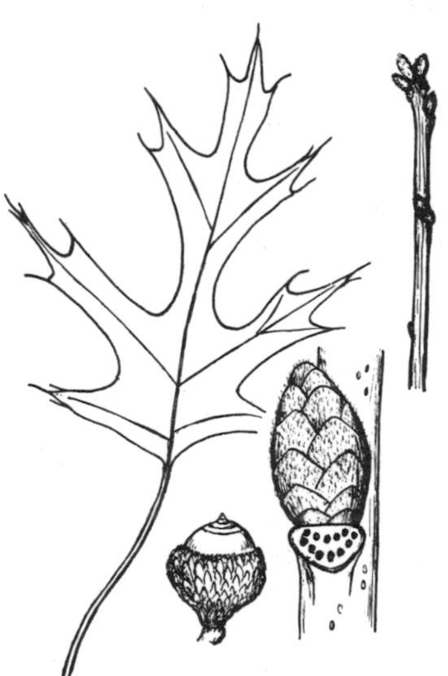
Fig. 74. Quercus coccinea

plain (Fig. 71).

9. Q. virginiana Mill. Live Oak. A broadly-crowned tree 15-25 m. high, with a trunk diameter of 6-9 dm; bark thick, dark, furrowed, finally flaky; buds small, 3 mm. long, ovoid, acute, remotely angled, the scales light brown, somewhat pubescent; acorn 1-2 cm. long, ovoid or subglobose, often striate; cups canescent, turbinate, about 1.5 cm. broad. Sandy soil, near the coast, Texas and Louisiana to Florida, and Virginia; also in Cuba, Mexico, and Central America.

10. Q. rubra L. Red Oak. (Q. borealis Michx.). A tree 20-30 m. high, 1-2 m. in diameter, the bark smooth, greenish-brown, on older stems broken into flat-topped ridges separated by narrow fissures; inner bark reddish; twigs glabrous; buds red, more or less hairy, about 5 mm. long; acorn 2-3 cm. long, narrowly ovoid or ellipsoid, the cup shallow, saucer-shaped. Upland woods, Prince Edward Island to Minnesota and Nebraska, south to Georgia and Oklahoma (Fig. 72).

11. Q. palustris Muenchh. Pin Oak. A tree 15-40 m. high, 6-9 dm. in diameter, easily recognized in winter by the drooping lower branches; twigs brown, glabrous; buds obtuse, brown, 3-4 mm. long; acorn globose or depressed, 1-1.5 cm. long, the cup flat, saucer-shaped, enclosing the nut only at the base. Bottomlands, Massachusetts to Michigan and Iowa, south to North Carolina, Louisiana, and Oklahoma (Fig. 73).

12. Q. coccinea Muenchh. Scarlet Oak. A tree 20-30 m. high, with a trunk diameter of 6-9 dm.; bark of the trunk rough, gray, the inner bark reddish; twigs glabrous; buds more or less silky, brownish-red, 5-6 mm. long; acorn 1.3-2 cm. long, subglobose or short-ovoid, usually with a few concentric rings about the apex; cup hemispherical, with a conical base. Dry soil, Maine to Ontario, south to Georgia, Mississippi, and Oklahoma (Fig. 74).

13. Q. shumardii Buckl. Shumard Oak. A large tree 30-45 m. high, 1-2 m. in diameter; bark on old trees very thick, broken into pale scaly ridges by deep darker colored furrows; buds 5-6 mm. long, ovoid, downy or glabrous; acorn 2-3 cm. long, oblong-ovoid; cup thick, shallow, the scales appressed. Bottomlands, most common on and near the coastal plain, Texas to Florida and Maryland, north in the Mississippi - Ohio Valley to Iowa and West Virginia (Fig. 75).

14. Q. velutina Lam. Black Oak. A tree 15-35 m. tall, the trunk 6-12 dm. in diameter; outer bark very dark brown, rough

Fig. 75. Quercus shumardii

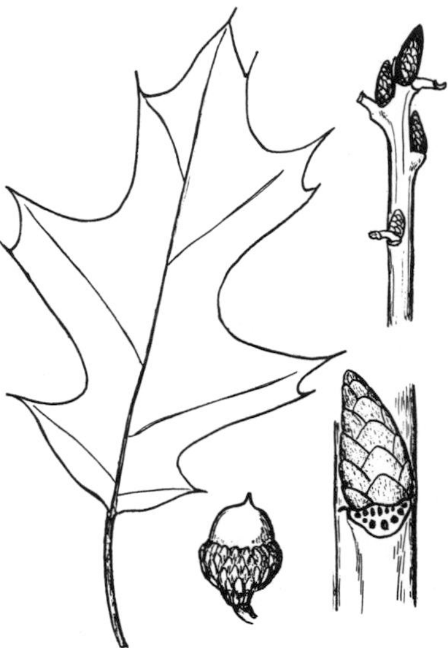

Fig. 76. Quercus velutina

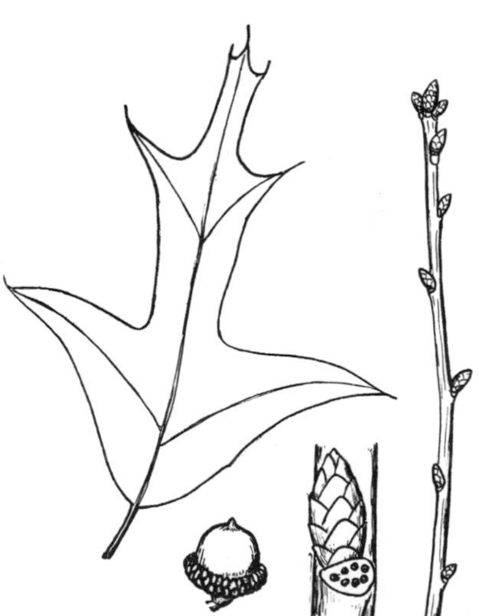

Fig. 77. Quercus falcata

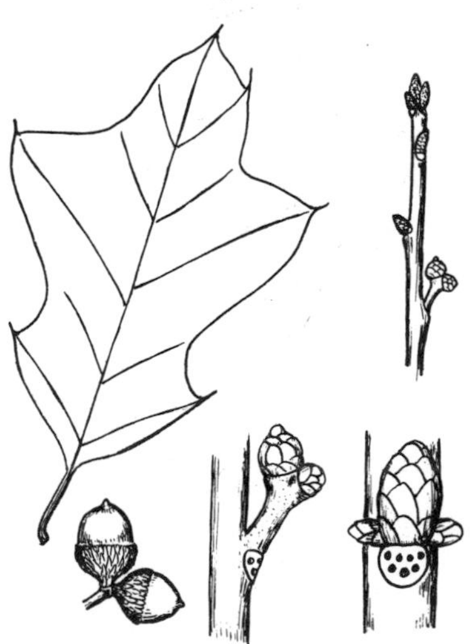

Fig. 78. Quercus ilicifolia

in low ridges, the inner bark bright orange-yellow; buds large, 7-10 mm. long-hairy, angular; acorn ovoid to hemispherical 1.2-2 cm. long, light brown, often pubescent; cup hemispherical. Dry woods, New Hampshire to Minnesota, south to Florida and Texas (Fig. 76).

15. Q. falcata Michx. Spanish Oak. Southern Red Oak. A tree 20-35 m. high, 6-9 dm. in diameter; bark black with broad scaly ridges; buds almost blood-red, more or less silky; acorn 1-1.5 cm. long, the nut enclosed only at the base by the saucer-shaped cup. Moist woods, most common on or near the coastal plain, Texas to Florida and Pennsylvania, north in the Mississippi-Ohio Valley to Missouri and West Virginia (Fig. 77).

16. Q. ilicifolia Wang. Scrub Oak. A straggling shrub or low tree 1-10 m. high, often forming dense thickets; twigs glabrous; buds 4 mm. long, glabrous; acorn globose-ovoid, 1-1.2 cm. long, the cup saucer-shaped. Barrens, Maine to New York, south mostly in the mountains to West Virginia and North Carolina (Fig. 78).

17. Q. marilandica Muenchh. Blackjack Oak. A tree 10-15 m. high, 3-4.5 dm. in diameter; bark black, very rough and blocky; twigs scurfy-puberulent; buds large, brown, angled; acorn ovoid, 1.5-2.5 cm. long; cup deep, the bracts appressed, pubescent. Dry soil, Florida to Texas, north to Maryland, New York, Illinois, and Nebraska (Fig. 79).

18. Q. imbricaria Michx. Shingle Oak. A tree 8-27 m. high, 3-9 dm. in diameter; bark with shallow fissures, the ridges with brown scales; twigs brown, glabrous; buds brown, 3 mm. long, pubescent; acorn subglobose, 1-1.4 cm. high; cup hemispheric, the bracts appressed. Bottomlands, New Jersey to Wisconsin and Nebraska, south to South Carolina and Kansas (Fig. 80).

ULMUS L. (Ulmaceae)

Deciduous trees. Twigs slender, zigzag, terete; pith small, round, continuous. Buds solitary or multiple, ovoid, obliquely sessile, the terminal lacking; scales about 6, 2-ranked. Leaf-scars alternate, 2-ranked, crescent-shaped or half-round; bundle-traces 3, or multiplied; stipule-scars unequal.

a. Older twigs often corky-winged, or with irregular corky outgrowths

b. Twigs often corky-winged — 4. U. alata
b. Twigs often with irregular corky ridges — 3. U. thomasi

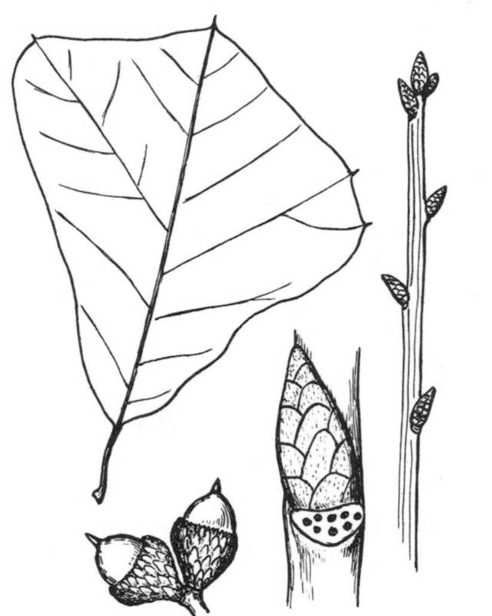

Fig. 79. Quercus marilandica

Fig. 80. Quercus imbricaria

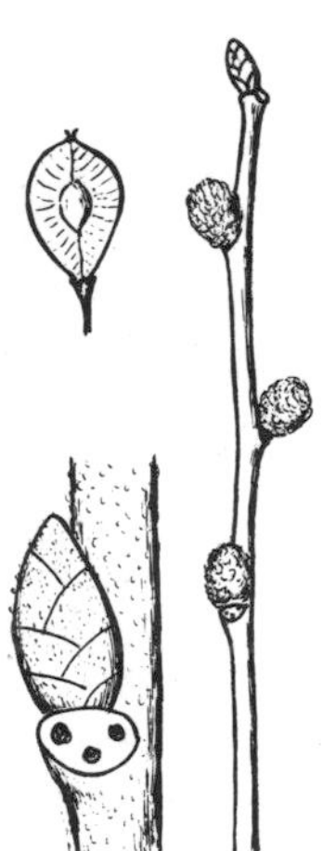

Fig. 81. Ulmus rubra

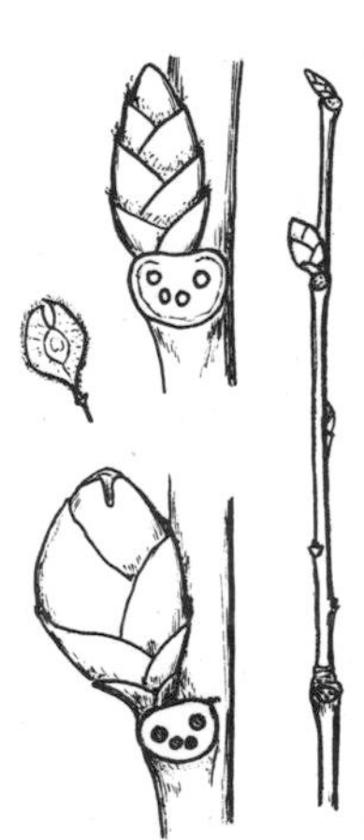

Fig. 82. Ulmus americana

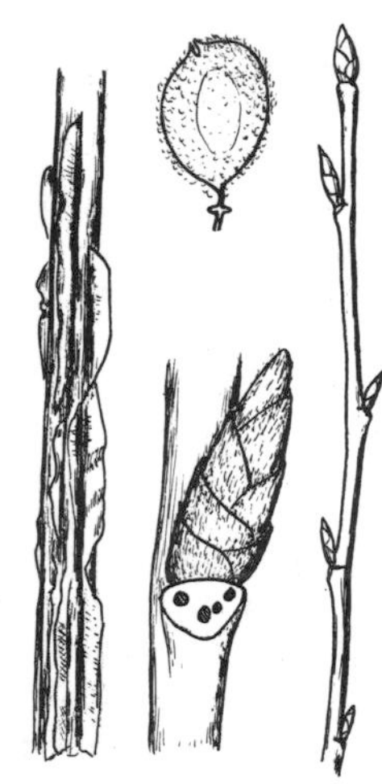

Fig. 83. Ulmus thomasi

a. Twigs without corky outgrowths

b. Buds downy, dark-colored, nearly black; twigs rough, ashy-gray in color 1. U. rubra

b. Buds brown, glabrous or pale-pubescent; twigs smooth or sparingly pilose 2. U. americana

1. U. rubra Muhl. Slippery Elm. Red Elm. (U. fulva Michx.). A tree 15-24 m. high, 3-7.5 dm. thick, with tough reddish wood; bark rough, gray, the inner bark very mucilaginous (whence the name, slippery elm); twigs gray-buff, rough-hairy; buds downy with red hairs; flowers nearly sessile, in groups of several, beginning to appear in March. Moist soil, New England to North Dakota, south to Florida and Texas (Fig. 81).

2. U. americana L. American Elm. A handsome tree, 20-50 m. high, 6-18 dm. in diameter, usually with a spreading vase-shaped crown and somewhat drooping branchlets; bark gray, flaky, with alternating light and dark layers in cross section; twigs glabrous or sparingly pubescent; buds brown, glabrous or somewhat pubescent; flowers long-pedicelled, in groups of 3 or 4, beginning to appear in March. Mostly in bottomlands, Quebec to Saskatchewan, south to Florida and Texas (Fig. 82).

3. U. thomasi Sarg. Cork Elm. Rock Elm (U. racemosa Thomas). A tree attaining a height of 35 m. and a diameter of 12 dm. twigs glabrous or puberulent; bud-scales downy-ciliate; branches often with corky ridges. Rich woods, mostly on limestone outcrops, New England to Minnesota and South Dakota, south to Tennessee and Kansas (Fig. 83).

4. U. alata Michx. Winged Elm. A small round-topped tree, up to 20 m. high; branches corky-winged; branchlets and buds nearly glabrous. Flowers in short racemes, appearing in March. Low woodlands, Florida to Texas, north to Virginia, Kentucky, and Missouri (Fig. 84).

CELTIS L. (Ulmaceae)

Deciduous trees or shrubs. Twigs rounded, slender; pith small, white, round, closely chambered, or continuous except at the nodes. Buds sessile, solitary, ovoid, with about 4 scales. Leaf-scars alternate, crescent-shaped or elliptical; bundle-traces 1 or 3; stipule-scars narrow. Fruit an ovoid or globose drupe, the mesocarp thin, pulpy, the endocarp bony.

a. Tree with buds 3-4 mm. long; drupes 8-11 mm. long 1. C. occidentalis

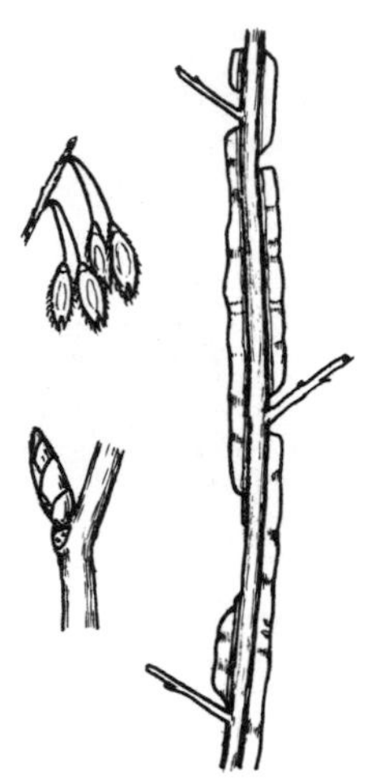

Fig. 84. Ulmus alata

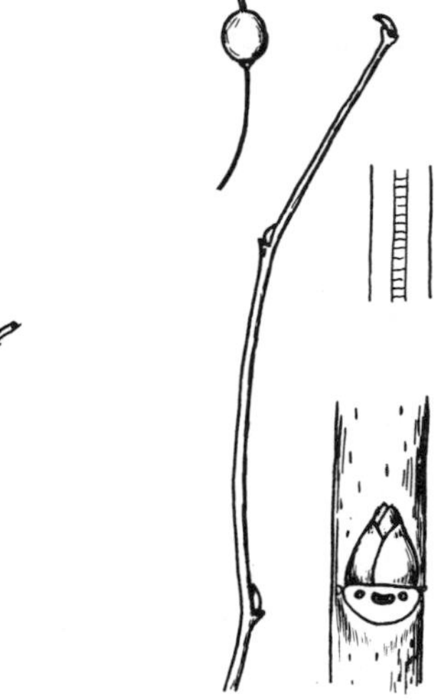

Fig. 85. Celtis occidentalis

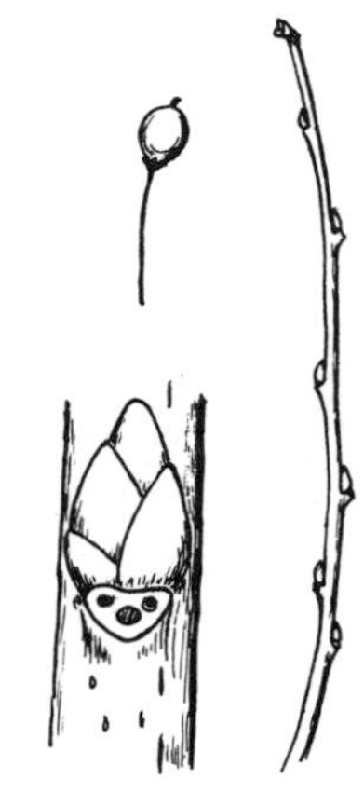

Fig. 86. Celtis pumila

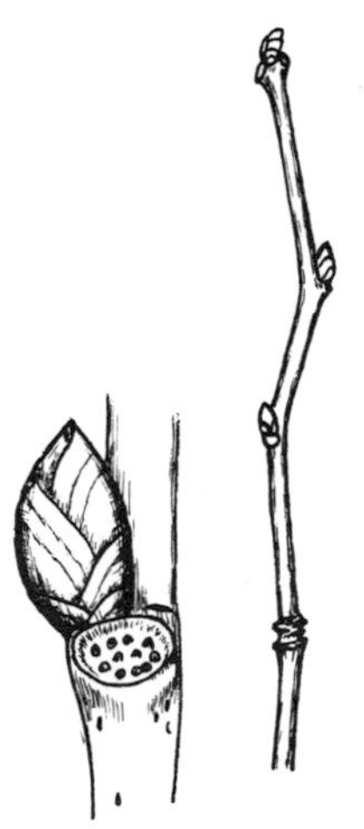

Fig. 87. Morus rubra

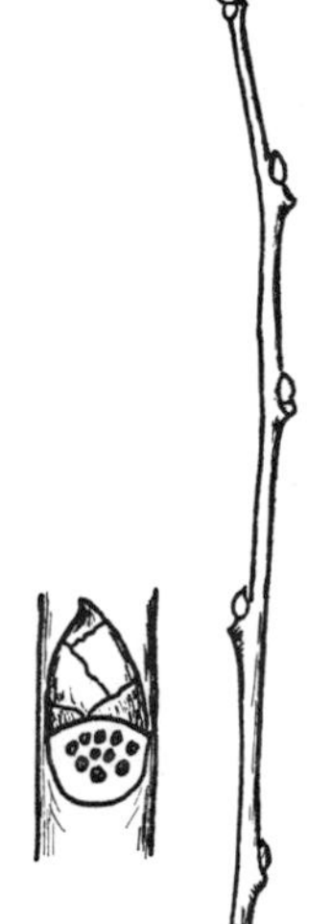

Fig. 88. Morus alba

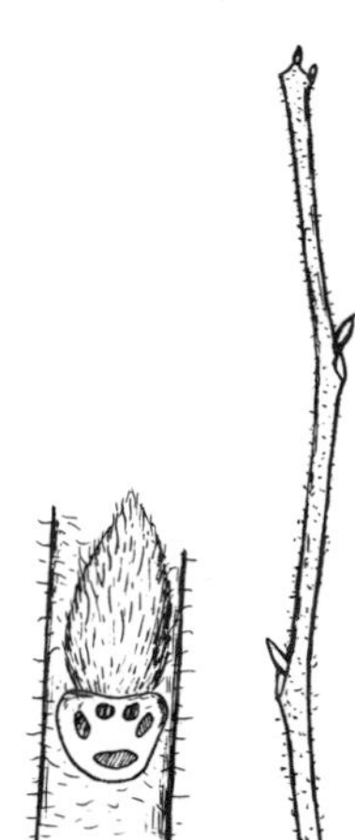

Fig. 89. Broussonetia papyrifera

a. Shrub with buds 1-2 mm. long; drupes 5-8 mm. long — 2. C. tenuifolia

1. C. occidentalis L. Hackberry. A tree 8-28 m. high, up to 7.5 cm. in diameter, the bark black, rough, corky-ridged or warty; buds 3-4 mm. long; drupes sweet and edible, 8-10 mm. in diameter, the flesh thin. Rich woods, Quebec to Idaho, south to Florida and Oklahoma (Fig. 85).

2. C. pumila Pursh. Dwarf Hackberry. A low straggling shrub, often fruiting copiously when but 6-9 dm. high; twigs brown; buds 1-2 mm. long; drupes globose. Dry slopes, Quebec to North Dakota, south to Georgia and Oklahoma (Fig. 86).

MORUS L. (Moraceae)

Deciduous trees with milky sap. Twigs moderate, rounded; pith moderate, round, continuous. Buds sessile, solitary or multiple, ovoid, oblique, with 3-6 scales; end-bud lacking. Leaf-scars alternate, round or half-round; bundle-traces numerous, scattered or in 3 groups; stipule-scars narrow.

a. Buds elongated, 6-8 mm. long, somewhat spreading — 1. M. rubra

a. Buds triangular-ovoid, short and closely appressed — 2. M. alba

1. M. rubra L. Red Mulberry. A tree 5-20 m. high, up to 2 m. in diameter; bark brown and rough; twigs often downy; buds spreading, 6-8 mm. long; bud-scales brown-margined. Rich woods, Ontario to Minnesota and South Dakota, south to Florida and Texas (Fig. 87).

2. M. alba L. White Mulberry. A small tree, sometimes 14 m. high and 1 m. in diameter; bark light gray, rough, the branches spreading; buds appressed, 3-4 mm. long; bud-scales generally uniformly colored. Introduced from Europe, much spread from cultivation and naturalized (Fig. 88).

BROUSSONETIA L'Her. (Moraceae)

Small deciduous trees with milky sap. Twigs moderate, rounded; pith large, white, with a thin green diaphragm at each node. Buds moderate, conical, solitary, sessile, the outer scale longitudinally striped. Leaf-scars alternate, 2-ranked, large, rounded; bundle-traces about 5; stipule-scars long and narrow.

1. B. papyrifera (L.) Vent. Paper Mulberry. A tree some-

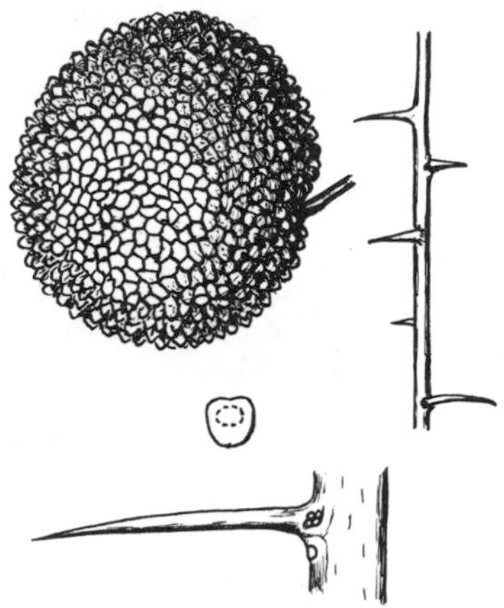

Fig. 90. Maclura pomifera

Fig. 91. Pyrularia pubera

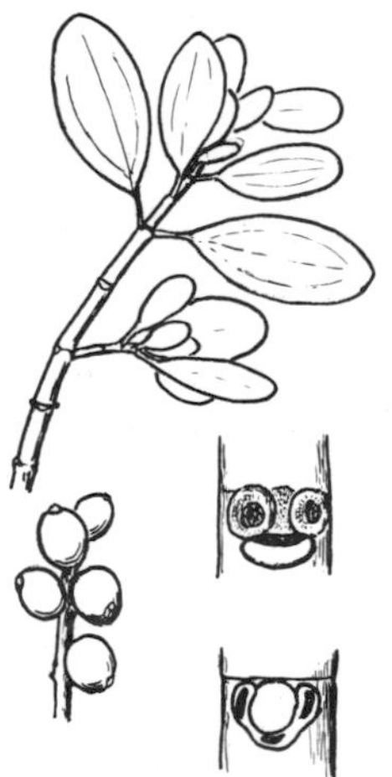

Fig. 92. Phoradendron flavescens

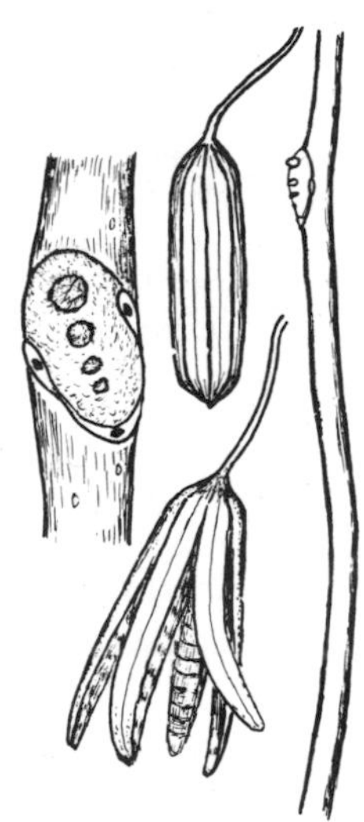

Fig. 93. Aristolochia durior

times 14 m. high, the young shoots hirsute-tomentose, relatively thick, greenish-gray. Introduced from Asia, spread from cultivation and naturalized (Fig. 89).

MACLURA Nutt. (Moraceae)

Deciduous trees with large axillary spines and milky juice. Twigs moderate, rounded, glabrous; pith moderate, round, pale, continuous. Buds small, depressed-globose, sessile, solitary or multiple, with about 5 scales; end-bud lacking. Leaf-scars alternate, half-round or triangular; bundle-traces several; stipule-scars small. Toxylon Raf.

1. M. pomifera (Raf.) Schneid. Osage-Orange. Hedge-Apple. A tree 10-20 m. high, with ridged brown bark and spreading branches; twigs buff or orange, glabrous; wood bright orange. Native in Texas, Oklahoma, and Arkansas, much cultivated and naturalized eastwards and northwards (Fig. 90).

PYRULARIA Michx. (Santalaceae)

Deciduous shrubs parasitic on roots of surrounding plants. Twigs rather large, terete, pubescent when young; pith large, rounded. Buds ovoid, brown. Leaf-scars alternate, rounded; bundle-traces 3; stipules none. In addition to the leaf-scars, there are conspicuous branch-scars, formed by self-pruning of young twigs, in which the bundle-traces are numerous, in an ellipse.

1. P. pubera Michx. Buffalonut. A straggling or erect much-branched shrub 1-4 m. high, with terete twigs. Rich woods, mountains of Pennsylvania and West Virginia to Georgia and Alabama (Fig. 91).

PHORADENDRON Nutt. (Loranthaceae)

Yellowish-green woody parasites on the branches of trees, with jointed, much-branched stems and thick firm persistent leaves. Fruit an ovoid or globose fleshy berry.

1. P. flavescens (Pursh) Nutt. American Mistletoe. A branching glabrous or slightly pubescent shrub, the twigs rather stout, terete, brittle at the base; leaves oblong or obovate, rounded at the apex, narrowed into short petioles, entire, 2.5-5 cm. long, 1-2 cm. wide, dark-green, leathery; berry globose, white, about 4 mm. in diameter. On various trees, Florida to Texas, north to New Jersey, West Virginia and Kansas (Fig. 92).

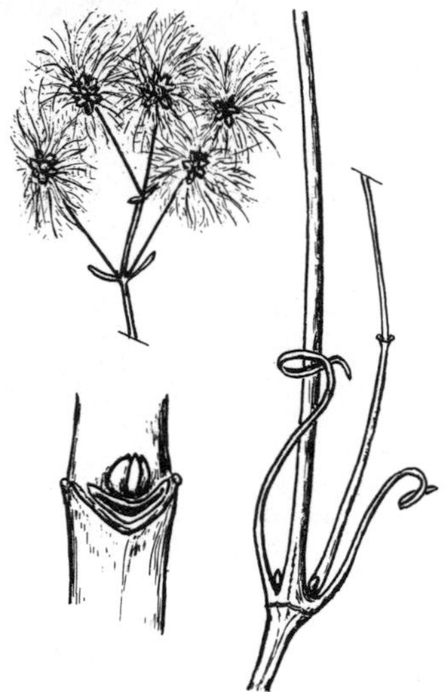

Fig. 94. Clematis virginiana

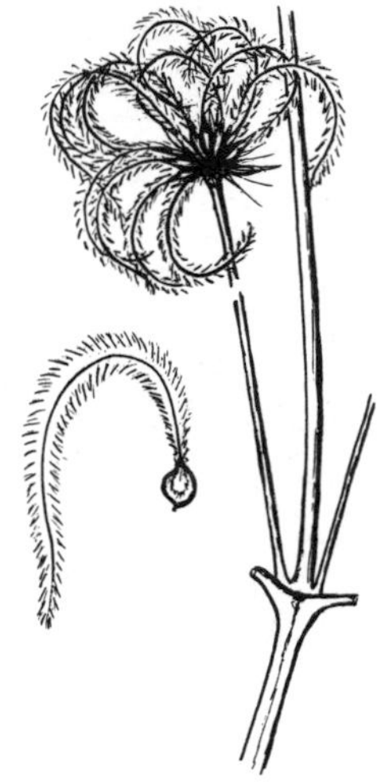

Fig. 95. Clematis viorna

Fig. 96. Clematis verticillaris

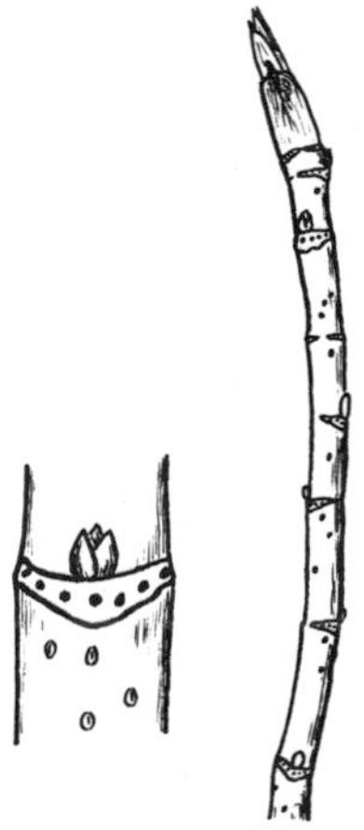

Fig. 97. Xanthorhiza simplicissima

ARISTOLOCHIA L. (Aristolochiaceae)

Soft-wooded deciduous shrubs, climbing high into trees. Stems terete, green, swollen at the nodes; pith large, rounded, continuous, pale. Buds small, sessile, rounded, superposed on a silky area, with 1 silky scale; terminal bud absent. Leaf-scars alternate, U-shaped; bundle-traces 3; stipule-scars lacking.

1. A. durior Hill. Dutchman's Pipe. Pipe vine. (A. macrophylla Lam.). A twining vine sometimes more than 10 m. high and 3 dm. in diameter, the branches slender, terete, green, glabrous; pith large. Rich woods, mountains of Pennsylvania and West Virginia to Georgia and Alabama (Fig. 93).

CLEMATIS L. (Ranunculaceae)

Soft-wooded plants, climbing by bending or clasping of the petioles. Stem 5- to 18-angled; pith angled or star-shaped, white, continuous, with firmer diaphragms at the nodes. Buds small, ovoid, or flattened, somewhat hairy, sessile, solitary, with 1-3 pairs of scales. Leaves opposite, withering but not falling, gradually torn off more or less piecemeal by winter winds.

a. Stems glaucous and glabrous — 3. C. verticillaris
a. Stems not glaucous, sometimes pubescent

b. Stems brown, finely pubescent; leaflets toothed — 1. C. virginiana
b. Stems red-brown, pubescent at the nodes; leaflets entire — 2. C. viorna

1. C. virginiana L. Virgin's Bower. A climbing vine, more or less woody; stem finely pubescent, brown, with about 6 distinct ridges, leaves normally 3-foliate, the leaflets ovate, toothed. Thickets, Quebec to Manitoba, south to Georgia, Louisiana, and Kansas (Fig. 94).

2. C. viorna L. Leatherflower. A vine climbing to a height of 3 m. or more; stem finely pubescent, red-brown; leaves mostly pinnate, firm, entire. Rich soil of thickets, Georgia to Texas, north to Pennsylvania and Iowa (Fig. 95).

3. C. verticillaris DC. Purple Virgin's Bower. A woody-stemmed climber, glaucous and glabrous; stem 6-sided, brown; leaves trifoliate. Rock slopes, Quebec to Wisconsin, south to West Virginia and Iowa (Fig. 96).

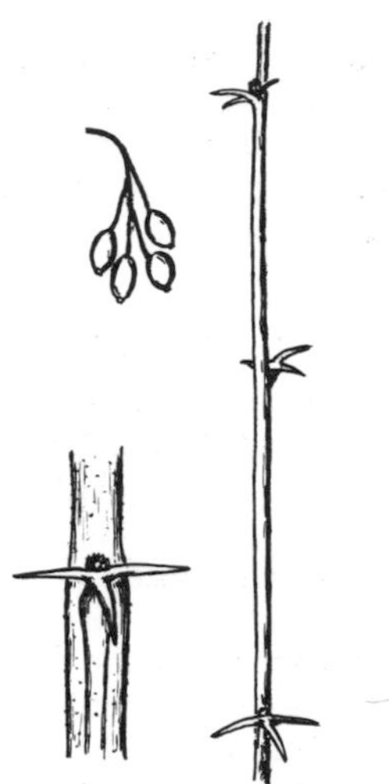

Fig. 98. Berberis canadensis

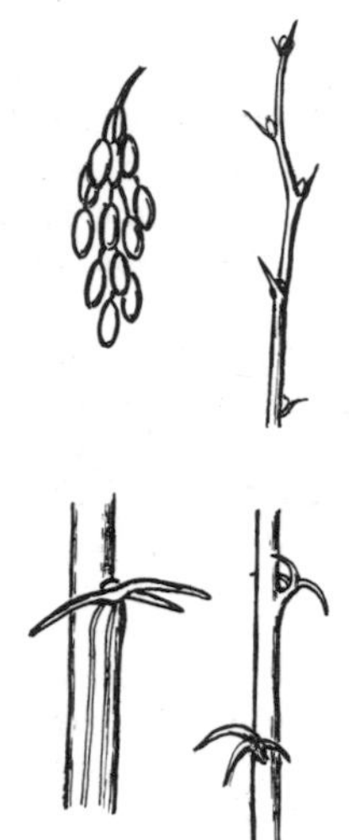

Fig. 99. Berberis vulgaris

Fig. 100. Berberis thunbergii

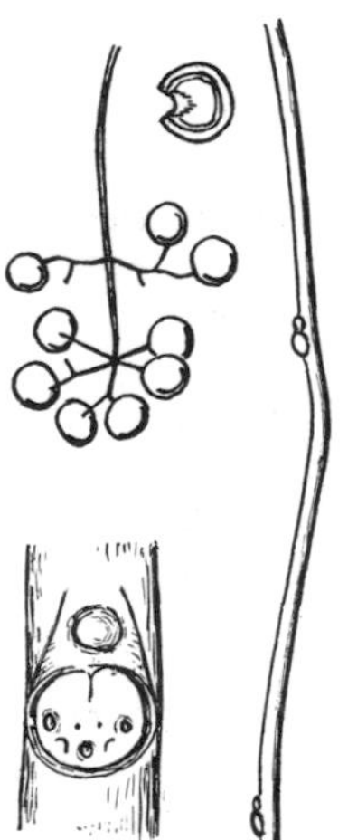

Fig. 101. Menispermum canadense

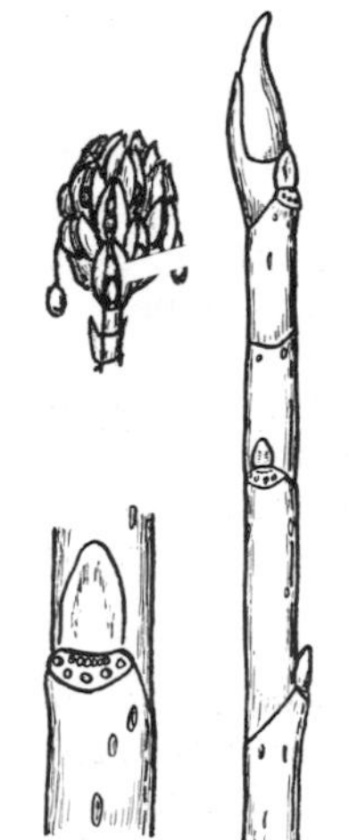

Fig. 102. Magnolia virginiana

Fig. 103. Magnolia acuminata

XANTHORHIZA Marsh. (Ranunculaceae)

Small, nearly unbranched shrubs, lemon-yellow when cut. Twigs terete, moderate, smooth; pith large, round, continuous. Lateral buds solitary, sessile, ovoid-oblong, compressed and flattened against the stem, with about 3 scales; terminal buds much larger, fusiform, terete, with about 5 scales. Leaf-scars alternate, low, slightly curved, more than half encircling the twig; bundle-traces about 11; stipule-scars none.

1. X. simplicissima Marsh. Shrubby Yellowroot. (X. apiifolia L'Her.). Stems clustered, 3-6 dm. high, glabrous; twigs gray; buds red-brown. Damp thickets, mostly in the mountains, New York to West Virginia, south to Alabama and Florida (Fig. 97).

BERBERIS L. (Berberidaceae)

Shrubs, usually with branched spines (representing modified leaves) subtending short spurs bearing fascicled foliage leaves. Wood and pith yellow. Twigs grooved, slender; pith relatively large, round, continuous. Buds rather small, solitary, sessile, ovoid, with about 6 scales. Leaf-scars alternate, small, half-round; bundle-traces 3, minute; stipule-scars none.

a. Twigs finely warty, slightly angled 1. B. canadensis
a. Twigs not warty

b. Twigs gray or buff 2. B. vulgaris
b. Twigs reddish-brown 3. B. thunbergii

1. B. canadensis Mill. American Barberry. A shrub 3-24 dm. high; twigs glabrous, slender, red-brown, warty; spines 3-parted, up to 12 mm. long; berries scarlet, oval or subglobose, 7-9 mm. long. Dry woodlands, mountains of West Virginia and Virginia to Georgia; also in Missouri; not native in Canada, despite the name (Fig. 98).

2. B. vulgaris L. European Barberry. A glabrous shrub 2-3 m. high, the branches arched and drooping at the ends, the twigs glabrous, gray or buff, strongly grooved; spines unbranched or mostly 3-parted; berries scarlet, ellipsoid. Introduced from Europe and abundantly naturalized in thickets (Fig. 99).

3. B. thunbergii DC. Japanese Barberry. Compact low-spreading, 0.5-1.5 m. high, with brown twigs and unbranched spines; berries ellipsoid to globose. Introduced from Asia, somewhat spreading, from cultivation (Fig. 100).

COCCULUS DC. (Menispermaceae)

Deciduous woody climbers. Twigs terete, slender; pith relatively large, continuous, white. Buds small, hairy. Leaf-scars alternate, elliptical; bundle-traces 3 or multiple; stipule-scars lacking. Drupe red, the stone crescent-shaped, with cross-ridges.

1. C. carolinus (L.) DC. Coralbeads. Twining to 4 m; branchlets pubescent; drupe 6-8 mm. long. Rich woods, Florida to Texas, north to Virginia, Kentucky, and Kansas.

MENISPERMUM L. (Menispermaceae)

Deciduous woody climbers. Twigs rounded, grooved, slender; pith large, continuous, white. Buds small, hairy. Leaf-scars alternate, elliptical, raised, concave; bundle-traces 3 to 7; stipule-scars none. Fruit a drupe, with a flattened ring-like or crescent-shaped stone, keeled on the back.

1. M. canadensis L. Moonseed. Stems 2-4 m. in length, twining over bushes; twigs green or buff, slightly pubescent or glabrous; drupe globose-oblong, 6-8 mm. in diameter before drying. Rich woods and thickets, Quebec to Manitoba, south to Georgia and Oklahoma (Fig. 101).

CALYCOCARPUM Nutt. (Menispermaceae)

Deciduous woody climbers. Twigs terete, slender, hairy or glabrescent; pith large, white, continuous. Leaf-scars alternate, elliptical; bundle-traces 3 or multiplied; stipule-scars none. Drupe with deep cup-like stone.

1. C. lyoni (Pursh) Gray. Cupseed. High climbing; drupe black, 2.5 cm. long. Rich soil, Florida to Louisiana, north to Kentucky, Missouri, and Kansas.

MAGNOLIA L. (Magnoliaceae)

Trees or shrubs with deciduous or evergreen foliage. Twigs moderate or thick, terete; pith rather large, continuous, round, often with firmer plates at intervals. Buds solitary, ovoid or fusiform, covered by a single scale, which, morphologically, is composed of the 2 coalescent stipules of the topmost leaf. Leaf-scars alternate, round or U-shaped; bundle-traces numerous, scattered; stipule-scars linear, encircling the twig.

a. Leaves more or less evergreen 1. M. virginiana

a. Leaves deciduous

b. Leaf-buds silky-hairy — 2. M. acuminata
b. Leaf-buds glabrous

c. Terminal buds about 2.5 cm. long — 4. M. fraseri
c. Terminal buds 3.5-5 cm. long — 3. M. tripetala

1. M. virginiana L. Sweet Bay. Small Magnolia. Large shrub or small tree up to 20 m. high, buds silky; branchlets glabrous or glabrate; leaves evergreen (or deciduous northwards), oval to broadly lanceolate, obtuse, glaucous, 0.8-1.5 dm. long. Swamps and wet woods, mostly in the coastal plain, Florida to Texas, north to Massachusetts, Tennessee, and Arkansas (Fig. 102).

2. M. acuminata L. Cucumber-tree. A tree 20-40 m. high, 6-12 dm. in diameter; bark grayish-brown, furrowed; twigs moderate; leaf-buds silky, 10-20 mm. long; leaf-scars U-shaped. The common name refers to the shape of the young fruit; only vestiges of the fruits are present in winter. Rich woods, New York and Ontario, south chiefly in the mountains, to Georgia and Arkansas (Fig. 103).

3. M. tripetala L. Umbrella Magnolia. A small tree 8-15 m. tall, 2.5-4 dm. thick; leaf-buds glabrous, purple, 2.5 cm. long, acute; leaf-scars large, oval. Rich woods, mostly in the mountains, Georgia to Arkansas, north to Pennsylvania and Missouri (Fig. 104).

4. M. fraseri Walt. Fraser Magnolia. A slender tree 9-20 m. tall, 3-4.5 dm. thick; bark smooth, dark brown; leaf-buds glabrous, purple, 3.5-5 cm. long; twigs glabrous and glaucous, relatively slender; leaf-scars rounded. Rich woods, mostly in the mountains, Virginia and West Virginia to Georgia and Alabama (Fig. 105).

LIRIODENDRON L. (Magnoliaceae)

Large deciduous trees. Twigs moderate, terete; pith rounded, pale, continuous, with firmer plates at intervals. Buds solitary or superposed, the lateral small, sessile, the terminal much larger, oblong, somewhat stalked, compressed, with 2 valvate scales (morphologically stipules of the topmost leaf, as in Magnolia). Leaf-scars alternate, large, round; bundle-traces many, in an irregular ellipse; stipule-scars linear, encircling the twig. Fruit, in the form of cone-like aggregates of samaras, present in winter.

1. L. tulipifera L. Tuliptree. Yellow-Poplar. A beautiful tree

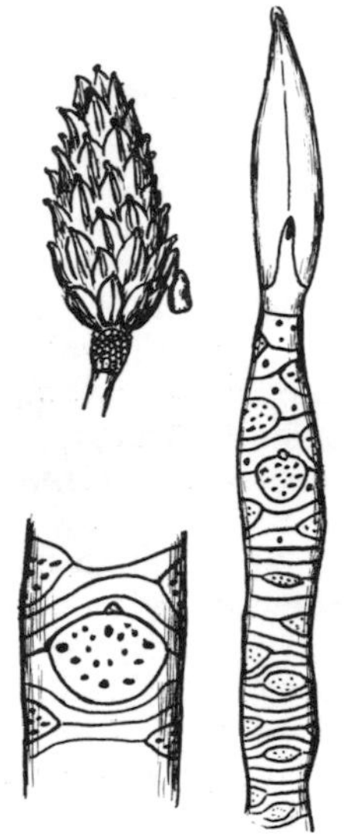

Fig. 104. Magnolia tripetala

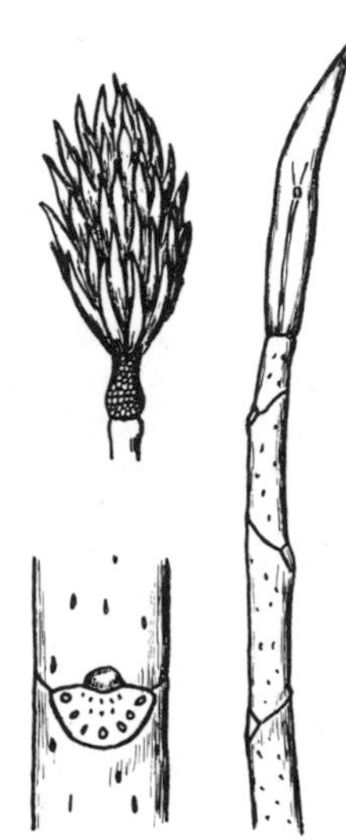

Fig. 105. Magnolia fraseri

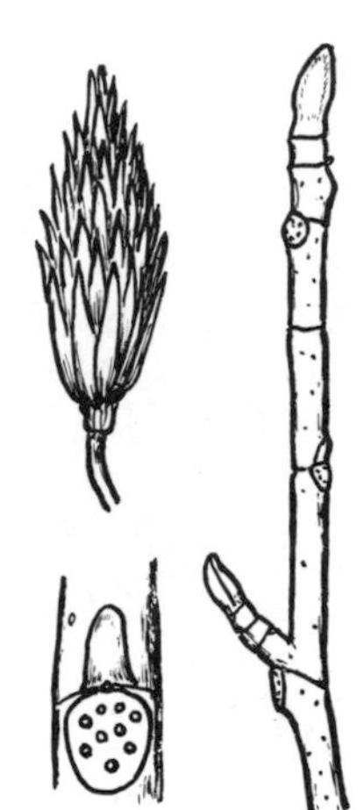

Fig. 106. Liriodendron tulipifera

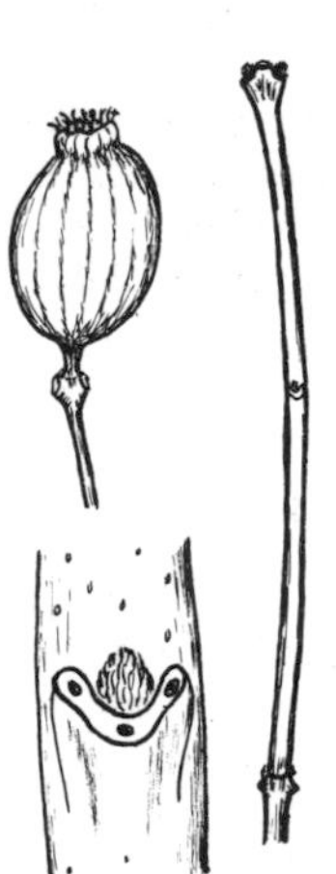

Fig. 107. Calycanthus fertilis

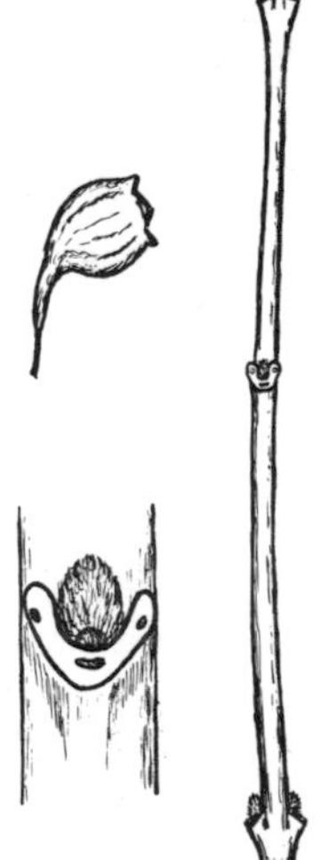

Fig. 108. Calycanthus floridus

Fig. 109. Asimina triloba

25-50 m. high, 1-3 m. in diameter; twigs red-brown; terminal buds 1.2 cm. long, shaped somewhat like a duck's bill; cone of fruit dry, oblong, acute, 7.5 cm. long. Rich soil, Massachusetts to Ontario and Michigan, south to Florida, Louisiana, and Arkansas (Fig. 106).

CALYCANTHUS L. (Calycanthaceae)

Aromatic deciduous shrubs, sparingly branched. Twigs moderate, compressed at the nodes; pith relatively large, somewhat 6-sided, white, continuous. Buds several in a single bud-like aggregate, sessile, round, brown-hairy, without evident scales. Leaf-scars opposite, horseshoe-shaped, raised; bundle-traces 3; stipule-scars none.

All parts of the plants and particularly the dried wood yield a pleasant fragrance, suggesting the common names Sweet-Scented Shrub, or Sweet Shrub (or sometimes merely "Shrub").

a. Twigs glabrescent or puberulous — 1. C. fertilis
a. Twigs more or less persistently villous — 2. C. floridus

1. C. fertilis Walt. Smooth Strawberry-Shrub. A branching shrub 12-27 dm. high, the branchlets glabrescent or slightly puberulous. Rich woods in the mountains, Georgia and Alabama, north to Pennsylvania (Fig. 107).

2. C. floridus L. Hairy Strawberry-Shrub. A branching shrub 6-27 dm. high, the branchlets more or less persistently villous. Rich woods, Florida to Mississippi, north to Virginia and West Virginia (Fig. 108).

ASIMINA Adans. (Annonaceae)

Deciduous shrubs or small trees. Twigs rounded, moderate; pith round, white, continuous, with firmer greenish plates at intervals in the second year's growth, often brown and chambered in old twigs. Terminal buds naked; lateral leaf-buds oblong, flower-buds globose. Leaf-scars alternate, half-round or V-shaped; bundle-traces 5 or 7, sometimes more; stipule-scars none.

1. A. triloba (L.) Dunal. Pawpaw. A tall shrub or small tree 3-12 m. high, the twigs slightly hairy or glabrescent. Rich woods, Florida to Texas, north to New York, Illinois, and Nebraska (Fig. 109).

SASSAFRAS Nees (Lauraceae)

Aromatic deciduous tree. Twigs green, glabrescent, rounded, moderate; pith moderate, white, continuous. Buds usually solitary, ovoid, sessile, subglobose; scales about 4. Leaf-scars alternate, small, half-round or crescent-shaped; bundle-trace a horizontal line more or less broken into 3; stipule-scars lacking. Fruit an oblong-globose blue drupe.

1. S. albidum (Nutt.) Nees. White Sassafras. A rough-barked tree 4-38 m. high, 3-9 dm. in diameter; twigs yellowish-green, glabrous, often glaucous; fruit a black drupe, persisting until early winter. Var. molle (Raf.) Fernald, Red Sassafras, has the twigs closely pubescent or puberulent. (S. variifolium (Salisb.) Ktze.; S. officinale Nees and Eberm.) Sassafras tea is prepared in winter from the bark of the root. Woods, Maine to Indiana and Iowa, south to Florida and Texas; often spreading aggressively and becoming weed-like (Fig. 110).

LINDERA Thunb. (Lauraceae)

Aromatic deciduous shrubs. Twigs rounded, slender, green; pith relatively large, round, white, continuous. Buds rather small, superposed; flower-buds collaterally multiple, green, globose, stalked; terminal bud absent. Leaf-scars alternate, crescent-shaped or half-round, small; bundle-traces 3, or sometimes confluent; stipule-scars none. Fruit an obovoid or oblong drupe. Benzoin Fabricius.

1. L. benzoin (L.) Blume. Spicebush. A shrub 2-5 m. high; twigs and buds glabrous; drupe red, 8-10 mm. high. Damp woods, Maine to Ontario and Iowa, south to Florida and Texas (Fig. 111).

PHILADELPHUS L. (Saxifragaceae)

Deciduous shrubs, mostly with exfoliating bark. Twigs more or less lined; pith moderate, rounded, pale, continuous. Buds solitary, sessile with 2 nearly valvate scales, end-bud lacking. Leaf-scars opposite, half-round, with a thin membrane partly covering the bud, bursting later in winter; bundle-traces 3; stipule-scars none. Fruit a capsule.

1. P. inodorus L. Mock-Orange. Shrub 2-3 m. high; twigs glabrous; fruit nearly solitary. Rocky slopes, Florida and Alabama to Virginia and Tennessee.

HYDRANGEA L. (Saxifragaceae)

Small soft-wooded deciduous shrubs. Twigs round, moderate to thick; pith relatively large, roundish, continuous, pale. Buds solitary, sessile or short-stalked, globose to oblong, with 4 to 6 scales. Leaf-scars opposite, crescent-shaped, rather large, often connected by lines around the stem; bundle-traces 3, 5, or 7; stipule-scars none. Fruit a membranous capsule, persistent in winter with the flat-topped inflorescence (cyme).

1. H. arborescens L. Wild Hydrangea. A bushy shrub 3-25 dm. high; twigs glabrous or nearly so; buds oblong, spreading. Rich woods, Florida to Louisiana and Oklahoma, north to New York and Missouri (Fig. 112).

RIBES L. (Saxifragaceae)

Deciduous sometimes spiny shrubs with shredding outer bark. Twigs rounded but decurrently ridged below the nodes, moderate; pith relatively large, pale, round, continuous but becoming porous in age. Buds small, solitary, sessile or short-stalked, ovoid or fusiform, with about 6 scales. Leaf-scars alternate, U-shaped or broadly crescent-shaped; bundle-traces 3; stipule-scars none.

- a. Stems reclining, with a strong skunk-like odor — 6. R. glandulosum
- a. Stems erect or arching
  - b. Buds ovoid, glandular or puberulent; leaf-scars rather broad
    - c. Bud-scales and twigs bearing rather conspicuous sessile resin-glands; buds glabrate — 8. R. americanum
    - c. Without resin-glands; buds gray-puberulent — 7. R. triste
  - b. Buds elongated-subfusiform; leaf-scars very narrow
    - c. Nodal prickles often large, up to 10 mm. long or more; buds glossy straw-colored — 2. R. missouriense
    - c. Prickles smaller; buds dull brown
      - d. Buds short, 3 mm. long, downy — 3. R. rotundifolium

Fig. 110. Sassafras albidum

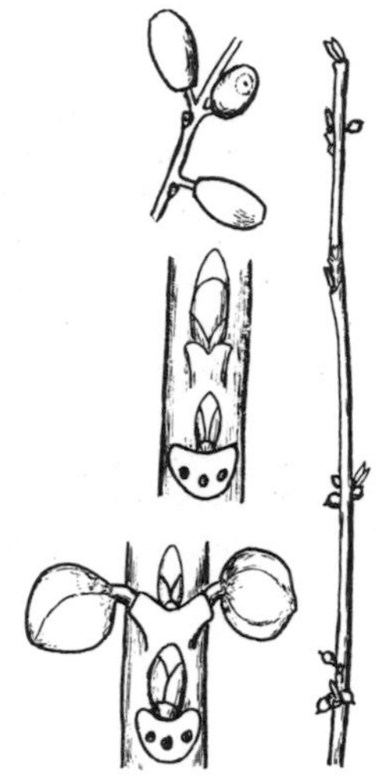

Fig. 111. Lindera benzoin

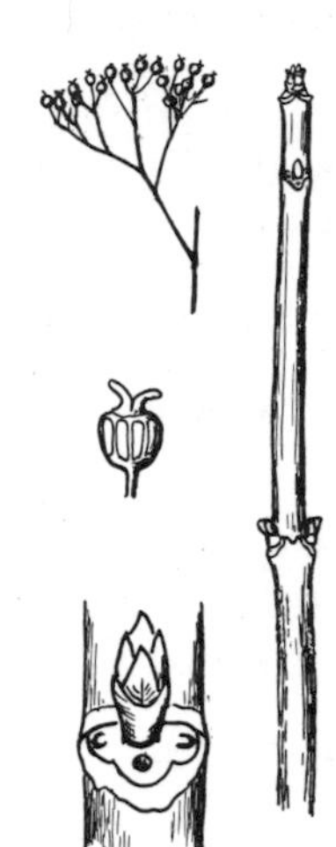

Fig. 112. Hydrangea arborescens

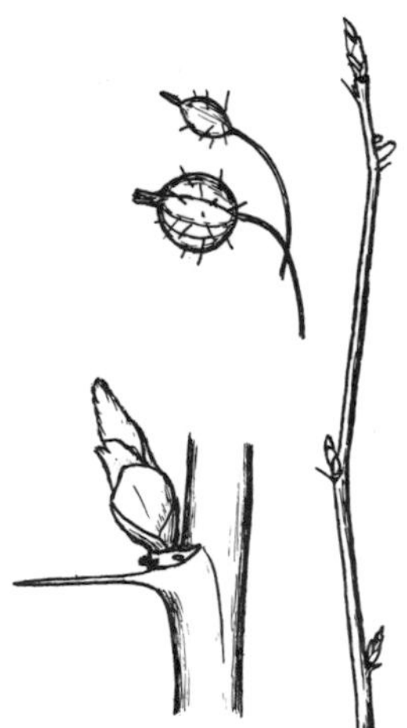

Fig. 113. Ribes cynosbati

Fig. 114. Ribes missouriense

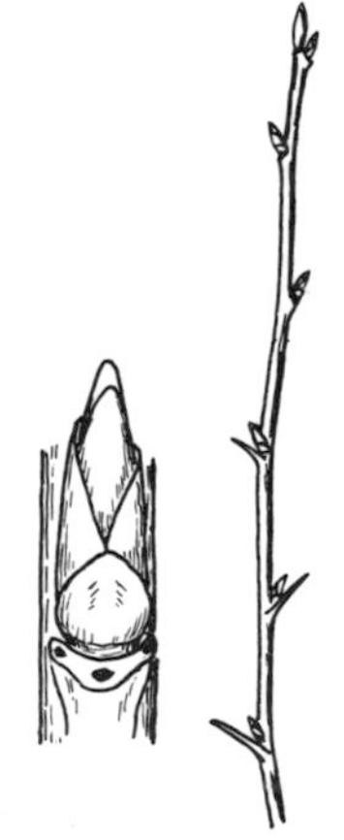

Fig. 115. Ribes rotundifolium

d. Buds rather long, 5-6 mm. long

e. Stems bristly and prickly, the nodal prickles about the length of the internodal bristles; twigs and buds glossy straw-colored 5. R. lacustre

e. Nodal prickles longer than the bristles

f. Twigs bristly and prickly, with quickly exfoliating epidermis; the fruiting canes mostly without bristles on the middle and upper internodes 4. R. hirtellum

f. Bark of internodes without bristles; nodal spines slender, 0. 5-1 cm. long 1. R. cynosbati

1. R. cynosbati L. Prickly Gooseberry. A shrub with erect or ascending branches; spines slender, 5-10 mm. long; bud-scales keeled, more or less silky; berries prickly (but not present in winter). Rocky woods, New Brunswick to Manitoba, south to North Carolina, Alabama and Missouri (Fig. 113).

2. R. missouriense Nutt. Missouri Gooseberry. An erect shrub to 2 m. high, bearing stout red spines 7-17 mm. long; buds elongated, fusiform, glossy, straw-colored; twigs grayish or whitish. Thickets, Connecticut to Minnesota and South Dakota, south to Kansas, Arkansas, Missouri and Tennessee (Fig. 114).

3. R. rotundifolium Michx. Smooth Gooseberry. A shrub with erect or spreading branches; spines only 2-5 mm. long, or sometimes lacking; buds 3 mm. long, downy; berry (absent in winter) not prickly. Rocky thickets in the mountains, Massachusetts to West Virginia and North Carolina (Fig. 115).

4. R. hirtellum Michx. Bristly Gooseberry. Shrub about 1 m. high; bark freely exfoliating, that of new canes prickly, of fruiting canes mostly without prickles on the lower internodes; nodal spines 3-8 mm. long. Rocky woods, Labrador to Manitoba, south to Pennsylvania, Illinois, and Minnesota (Fig. 116).

5. R. lacustre (Pers.) Poir. Black Currant. Shrub about 1 m. high; young stems clothed with bristly prickles and with weak

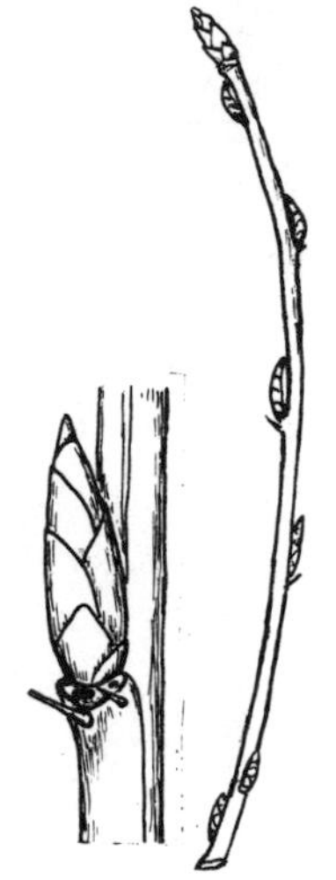

Fig. 116. Ribes hirtellum

Fig. 117. Ribes lacustre

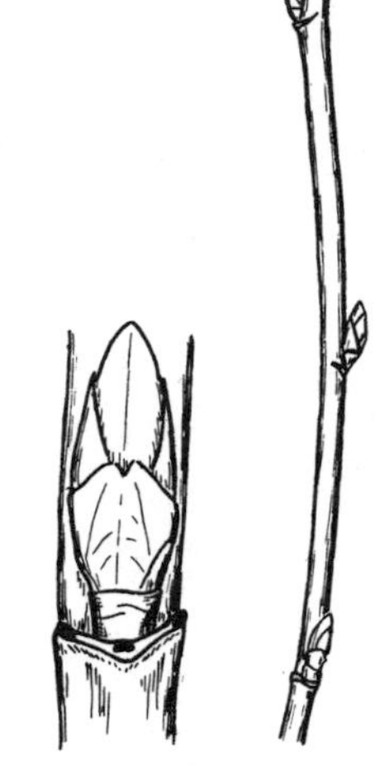

Fig. 118. Ribes glandulosum

Fig. 119. Ribes triste

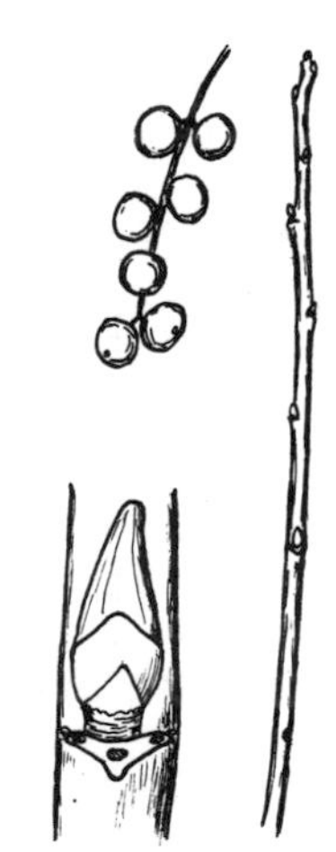

Fig. 120. Ribes americanum

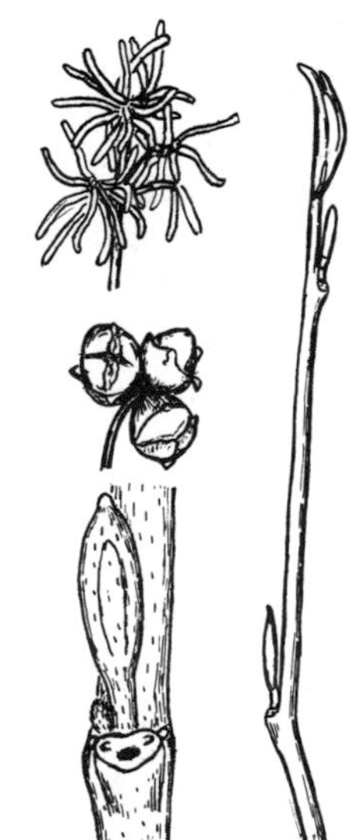

Fig. 121. Hamamelis virginiana

thorns, the nodal spines about the length of the internodal; twigs and buds glossy straw-colored. Cold damp woods, Newfoundland to Alaska, south in the eastern mountains to Tennessee, in the west to Colorado, Utah, and California (Fig. 117).

6. R. glandulosum Grauer. Skunk Currant. (R. prostratum L'Her.). A decumbent or spreading very ill-scented shrub; bark blackish; twigs unarmed, glabrate. Wet woods, Labrador to Mackenzie and British Columbia, south to North Carolina, Michigan and Saskatchewan (Fig. 118).

7. R. triste Pall. Red Currant. Straggling or reclining shrub, the branches often freely rooting; twigs quickly glabrate. Cool woods and swamps, Labrador to Alaska, south to Pennsylvania, Michigan, Minnesota, and Oregon (Fig. 119).

8. R. americanum Mill. Wild Black Currant. An erect unarmed shrub 1.5 m. or less high, with spreading branches;buds ovoid; bud-scales and twigs glabrate, bearing large conspicuous resin-glands; leaf-scars broad. Rich slopes, New Brunswick to Alberta, south to Virginia, Missouri, and New Mexico (Fig. 120).

HAMAMELIS L. (Hamamelidaceae)

Deciduous shrubs or small trees. Twigs rounded, zigzag, slender, stellate-tomentose to glabrate; pith small, round, green, continuous. Buds moderate, stalked, oblong, tomentose. Leaf-scars alternate, 2-ranked, half-round or 3-lobed; bundle-traces 3, often compound; stipule-scars unequal, one round, the other elongated. The leaves have a curious double abscission layer, the petioles falling normally in autumn and the surface of the leaf-scar again abscissing in spring.

1. H. virginiana L. Witch-hazel. A shrub or small tree 5-8 m. high, the stems 10-25 cm. in diameter; buds sordid-yellow, 5-8 mm. long, including the stalk. The yellow flowers appear in autumn and the obovoid-pubescent capsules, 1-1.5 cm. long, mature a year later. Woods, Quebec to Minnesota, south to Georgia and Louisiana (Fig. 121).

LIQUIDAMBAR L. (Hamamelidaceae)

Conical deciduous trees, with a resinous sap. Twigs moderate, terete and smooth or with variously developed corky ridges; pith angled, continuous or nearly so, white or brownish. Buds solitary, ovoid, with about 6 exposed scales. Leaf-scars alternate, elliptical or triangular; bundle-traces 3, large; stipule-scars none.

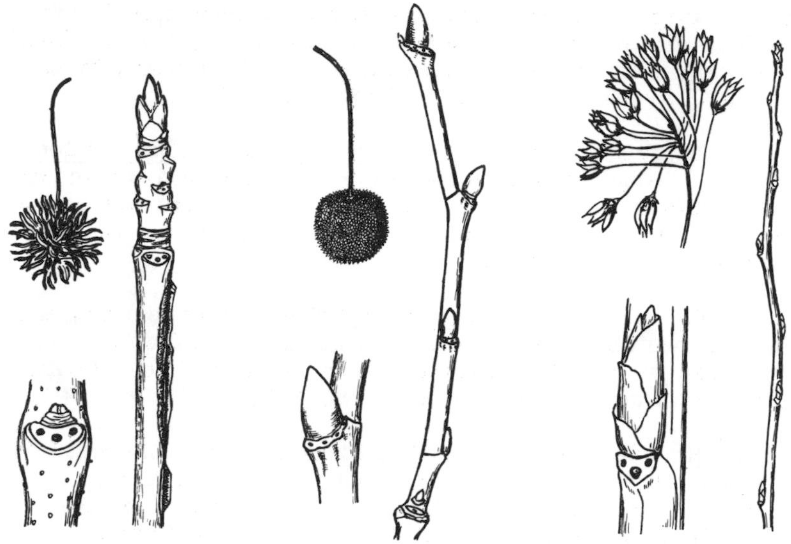

Fig. 122. Liquidambar styraciflua

Fig. 123. Platánus occidentalis

Fig. 124. Physocarpus opulifolius

Fig. 125. Spiraea alba

Fig. 126. Spiraea latifolia

Fig. 127. Spiraea tomentosa

Fruit a long-stalked globose dry head formed by numerous somewhat cohering capsules.

1. L. styraciflua L. Sweetgum. A tree 15-35 m. high, the trunk 6-12 dm. in diameter, the bark very rough; branches often winged with corky ridges; twigs shiny, green or brown, aromatic; buds 6-12 mm. long, the scales orange-brown. Moist woods, Florida to Texas, Mexico and Central America, north to New York, Ohio and Missouri (Fig. 122).

PLATANUS L. (Platanaceae)

Large open deciduous trees with exfoliating bark. Twigs moderate, rounded, glabrous, buff, zigzag; pith moderate, white or brownish, rounded, continuous. Buds solitary, rather large, sessile, conical, with a single cap-like scale; end bud lacking. Leaf-scars alternate 2-ranked, ring-like and nearly encircling the buds; bundle-traces 5, 7, or 9, large; stipule-scars narrow, encircling the twig. Fruit a globose head of elongated obovoid achenes, each with a circle of upright brown hairs at the base.

1. P. occidentalis L. Sycamore. Buttonwood. Plane-tree. A tree 35-50 m. high, 1-3 m. in diameter; bark covered with broad curling scales which are shed off, exposing the smooth greenish-white surface; fruit-balls solitary, on long stalks, present in winter, but disintegrating as the season progresses. Rich moist soil, Maine to Ontario and Nebraska, south to Florida and Texas. The largest tree of the northeast (Fig. 123).

PHYSOCARPUS Maxim. (Rosaceae)

Loosely branching deciduous shrubs with quickly shredding brown bark. Twigs terete, 5-lined at the nodes, slender; pith large, brownish, round, continuous. Buds small, solitary, sessile, conical-oblong or ovoid, with about 5 loose scales. Leaf-scars alternate, elliptical or 3-lobed; bundle-traces 3, unequal, the lower one larger than the others; stipule-scars small.

1. P. opulifolius (L.) Maxim. Ninebark. A shrub 1-3 m. high, with long branches, the old bark loose and separating in numerous thin layers (whence the common name); buds pointed, appressed; twigs glabrous; fruit, as clustered small dry follicles, present in winter. Shores and rocky banks, Quebec to Ontario, Minnesota and Colorado, south to South Carolina and Arkansas (Fig. 124).

SPIRAEA L. (Rosaceae)

Low deciduous shrubs, usually with stiff wand-like branches. Twigs round or somewhat angled near the nodes, slender; pith

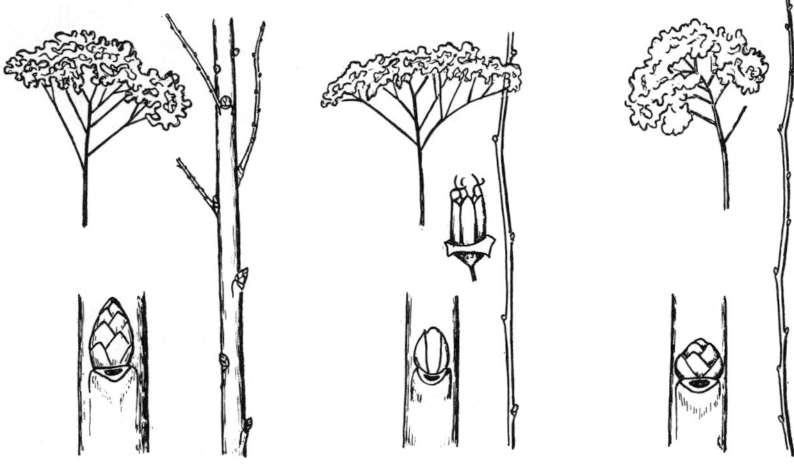

Fig. 128. Spiraea japonica

Fig. 129. Spiraea corymbosa

Fig. 130. Spiraea virginiana

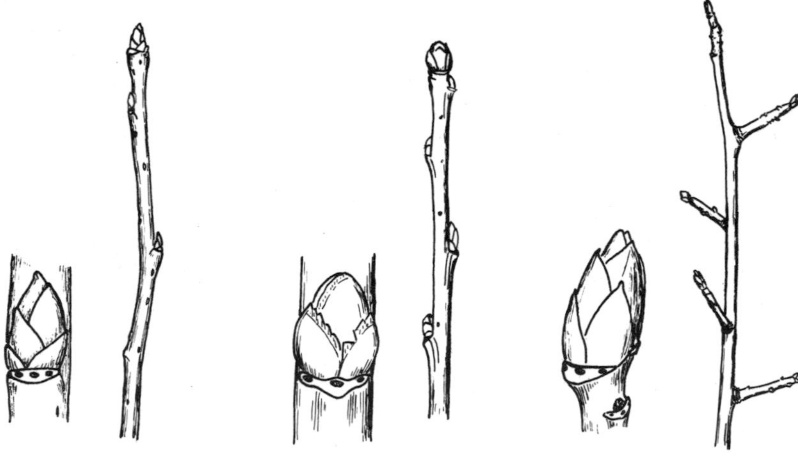

Fig. 131. Pyrus communis

Fig. 132. Pyrus malus

Fig. 133. Pyrus coronaria

small, round, continuous. Buds small, solitary or collaterally multiple, sessile, globose to fusiform, with about 6 scales. Leaf-scars alternate, half-round or crescent-shaped, minute; bundle-trace 1. Inflorescence paniculate or corymbose, persistent through the winter.

a. Inflorescence-vestiges paniculate

  b. Stems puberulent or glabrous

    c. Branchlets of panicle puberulent or tomentulose — 1. S. alba
    c. Branchlets of panicle glabrous — 2. S. latifolia

  b. Stems very woolly — 3. S. tomentosa

a. Inflorescence-vestiges corymbose

  b. Stems pubescent — 4. S. japonica
  b. Stems glabrous

    c. Stems simple or nearly so — 5. S. corymbosa
    c. Stems much-branched — 6. S. virginiana

1. *S. alba* Du Roi. *Meadowsweet. Wild Spiraea. Pipestem.* Erect shrub, 0.3-2 m. high; twigs tough, yellowish-brown, more or less angled, pubescent, at least in the paniculate inflorescence. Low ground, Quebec and Vermont to Saskatchewan, south to Missouri, Ohio, and in the mountains to North Carolina (Fig. 125).

2. *S. latifolia* (Ait.) Borkh. *Meadowsweet.* Erect shrub 3-12 dm. high, with tough yellowish-brown stems; panicles mostly open-pyramidal, 0.5-3 dm. long, the branchlets glabrous. Low grounds, Newfoundland to Michigan, south in the mountains to North Carolina (Fig. 126).

3. *S. tomentosa* L. *Hardhack. Steeplebush.* Stems angled, 1-2 m. high, pubescent with rusty wool; buds solitary, ovoid, short, with several exposed scales. Low grounds, Prince Edward Island to Manitoba, south to Georgia and Arkansas (Fig. 127).

4. *S. japonica* L. f. *Japanese Spiraea.* Stems 1.5 m. high or less, with gray or dingy inflorescence; twigs terete; inflorescence corymbose. Introduced from Asia, spreading from cultivation into thickets and somewhat naturalized (Fig. 128).

5. *S. corymbosa* Raf. *Corymbed Spiraea.* Stems erect, less than 1 m. high; twigs terete, glabrous, bright red-brown; buds

ovoid, solitary, with several exposed scales; inflorescence corymbose. Rocky banks, in the mountains, New Jersey south to Georgia and Kentucky (Fig. 129).

6. S. virginiana Britt. Virginia Spiraea. Stem much-branched, to 1.2 m. high; twigs often glaucous, glabrous or pubescent, more or less angled, inflorescence corymbose. Rocky places, in the mountains, West Virginia to Tennessee (Fig. 130).

PYRUS L. (Rosaceae)

Deciduous shrubs or medium-sized trees. Twigs moderate, round or slightly angled, sometimes ending in sharp points; pith somewhat angled, continuous. Buds moderate, solitary, sessile, with about 4 scales. Leaf-scars alternate, linear; bundle-traces 3; stipule scars none.

a. Twigs and buds glabrous; bud-scales submucronate, not margined; twigs olive — 1. P. communis
a. Buds pubescent, their scales sometimes margined

b. Buds blunt-ovoid; scales subobtuse; unarmed — 2. P. malus
b. Buds conical-oblong; scales acute; and twigs sharp-pointed — 3. P. coronaria

1. P. communis L. Pear. Small tree, to about 15 m. high, when wild often with spinescent branches; branchlets olive, glabrous or glabrate, with inconspicuous lenticels. Introduced from Eurasia, spreading slightly from cultivation into thickets (Fig. 131).

2. P. malus L. Apple. (Malus pumila Mill.). A tree to 15 m. tall, with spreading branches, the trunk sometimes 1 m. in diameter; twigs more or less pubescent; buds blunt-ovoid, pubescent; scales more or less obtuse. Introduced from Eurasia, spreading and naturalized (Fig. 132).

3. P. coronaria L. Wild Crabapple. (Malus coronaria Mill.). Tree, somewhat armed (at least bearing sharp-pointed twigs), 6-10 m. high, the trunk 2.5-3.5 dm. in diameter; twigs glabrate; bundle-traces 3; buds conical-oblong, scales pubescent, acute. Thickets, New York and Ontario to Minnesota, south to Kansas and North Carolina (Fig. 133).

ARONIA Medic. (Rosaceae)

Deciduous shrubs. Twigs moderate, rounded, glabrous; pith moderate, rounded, continuous. Buds solitary, sessile, oblong, flattened, appressed. Leaf-scars alternate, crescent-shaped or U-shaped; bundle-traces 3; stipule-scars none.

a. Twigs and buds somewhat woolly ..... 1. A. arbutifolia
a. Twigs and buds glabrous ..... 2. A. melanocarpa

1. A. arbutifolia (L.) Ell. Red Chokeberry. (Pyrus arbutifolia (L.) L.f.). A shrub 1-2.5 m. high; twigs and buds more or less woolly; leaf-scars low; bundle-traces 3; fruits red, persistent into winter. Swamps and damp barrens, Ontario to Michigan and Missouri, south to Florida and Texas (Fig. 134).

2. A. melanocarpa (Michx.) Ell. Black Chokeberry. (Pyrus melanocarpa (Michx.) Willd.) A shrub up to 4 m. high, but generally considerably lower; twigs and buds glabrous; leaf-scars low; bundle-traces 3; fruits black, persistent into winter. Moist thickets and barrens, Newfoundland to Minnesota, south to South Carolina and Tennessee (Fig. 135).

SORBUS L. (Rosaceae)

Erect-branched small or moderate deciduous trees. Twigs moderate, with prominent lenticels, rounded; pith round, brown, continuous. Buds oblong, the terminal large, the lateral reduced, solitary, sessile, more or less pubescent with long hairs matted in gum. Leaf-scars alternate, raised, crescent-shaped or linear; bundle-traces 3 or 5; stipule-scars none.

1. S. americana Marsh. Mountain-Ash. (Pyrus americana (Marsh.) DC.). A tree 6-10 m. tall, 2-3 dm. in diameter; twigs rather thick, with prominent lenticels; buds oblong, the terminal large, the lateral smaller, solitary, sessile, more or less pubescent with long hairs matted in gum; leaf-scars raised, crescent-shaped or linear; bundle-traces 3 or 5; fruits orange-red, 4-6 mm. in diameter, holding through the winter, but fading. Woods, Newfoundland to Minnesota, south to Georgia and Illinois (Fig. 136).

AMELANCHIER Medic. (Rosaceae)

Deciduous trees or shrubs. Twigs rather slender, zigzag, terete; pith more or less 5-angled, continuous, pale. Buds moderate, sessile, solitary, elongated, with about 6 often twisted

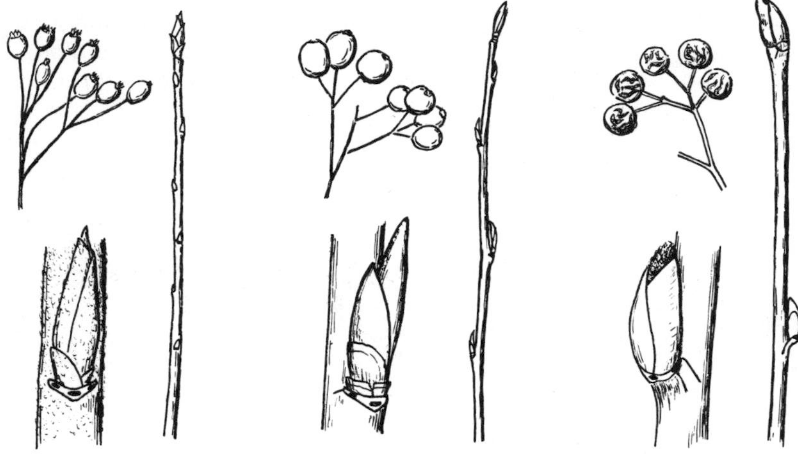

Fig. 134. Aronia arbutifolia

Fig. 135. Aronia melanocarpa

Fig. 136. Sorbus americana

Fig. 137. Amelanchier sanguinea

Fig. 138. Amelanchier humilis

Fig. 139. Amelanchier arborea

scales. Leaf-scars alternate, crescent-shaped or U-shaped; bundle-traces 3; stipule-scars none.

a. Low shrubs, 0.3-8 m. high

b. Stoloniferous, the individual stems scattered or loosely colonial — 1. A. humilis

b. Not stoloniferous

c. Straggling, slender, often arching shrubs, the stems solitary or few in a clump — 2. A. sanguinea

c. Stems several together, loosely caespitose-fastigiate; bog plants — 5. A. bartramiana

a. Trees, up to 20 m. high

b. Leaves densely pubescent in the bud — 3. A. arborea

b. Leaves glabrous or only slightly pubescent in the bud — 4. A. laevis

1. *A. humilis* Wieg. *Low Serviceberry.* A stiffly upright shrub or small tree 0.3-8 m. high, growing in patches from rhizome-like bases (stoloniferous), the individual stems scattered; buds 4-9 mm. long, dull. Rocky banks, often calcareous, Quebec to Minnesota and South Dakota, south to West Virginia and Ohio (Fig. 137).

2. *A. sanguinea* Pursh. *Roundleaf Serviceberry.* A straggling or arching slender shrub 1-2.5 m. high, 3 cm. or less in diameter, not stoloniferous and not forming colonies, the stems solitary or few in a clump; branchlets red or reddish-brown; buds slender, reddish-brown, 6-7 mm. long, dull. Open woods or rocky slopes, Quebec and Ontario, south to Iowa, Michigan, and North Carolina (Fig. 138).

3. *A. arborea* (Michx.) Fernald. *Juneberry. Common Serviceberry. Downy Serviceberry.* "Sarvis". (*A canadensis* of authors, not Medic). An irregular bushy tree 5-20 m. high, the trunk up to 4 dm. in diameter; leaves densely pubescent in the bud; buds 6-13 mm. long. Rich woods, New Brunswick to Minnesota, south to Florida, Louisiana, and Oklahoma. Our most common and widely distributed species (Fig. 139).

4. A. laevis Wieg. Smooth Serviceberry. A tree 13 m. high or less; leaves sparingly pubescent or glabrous in the bud; buds 0.9-1.7 cm. long. Thickets, Newfoundland to Ontario, south to Iowa and Illinois and in the mountains to Georgia (Fig. 140).

5. A. bartramiana (Tausch) Roemer. Glade Serviceberry. (A. oligocarpa (Michx.) Roemer). A shrub 0.5-2.5 m. high, the stems several together, loosely caespitose-fastigiate; leaves imbricate in the bud. Bogs and moist slopes, Labrador to Ontario and Minnesota, south in the mountains to West Virginia (Fig. 141).

CRATAEGUS L.*Hawthorns. Red Haws (Rosaceae)

Deciduous shrubs or trees, usually with well-developed twig-spines. Twigs moderate or rather slender, terete, pith rather small, continuous, round. Buds solitary or collaterally branched, sessile, round or oblong-ovoid, with about 6 exposed scales. Leaf-scars alternate, narrowly crescent-shaped; bundle-traces 3; stipule-scars small.

Crataegus is a large and exceptionally complicated genus, with more than 100 species recognized from the range of this manual. The plants, however, are very hard to identify, even from material having leaves, flowers, and fruit, and it is a hopeless task to attempt to construct a key to distinguish them in the winter state. In the present treatment, nevertheless, an effort has been made to provide winter characters for the different groups (Series) of the region. Even this has proved very difficult and the following key must be regarded as quite provisional.

a. Slender shrubs .8-1.5 m. tall with few simple or little-branched stems.

Series 2. Parvifoliae (C. uniflora Muench.)

a. Arborescent shrubs or trees with stout boles or bifurcating branches.

b. Usually shrubby but sometimes becoming small trees under favorable conditions.

c. Branches mostly ascending forming a narrow crown; thorns numerous, long and stout.

Series 14. Macracanthae. (C. succulenta Link)

c. Lower branches spreading, forming rounded crowns usually as

* Based on data furnished by Ernest J. Palmer

broad as high, armed with slender thorns or sometimes nearly thornless.

d. Bark comparatively thin and smooth, exfoliating in small thin flakes.

Series 1. Microcarpae. (C. spathulata Michx.,etc.)

d. Bark on old boles or branches comparatively thick, more or less fissured or scaly.

e. Thorns usually straight and slender, 2-3.5 cm. long. No compound thorns.

Series 4. Pulcherrimae. (C. ancisa Beadle, etc.)

e. Thorns stouter, often curved, 3-5 cm. long. Compound thorns sometimes on boles and larger branches.

Series 3. Intricatae. (C. intricata Lange, etc.)

b. Usually arborescent, but sometimes appearing shrubby in young state.

c. Boles and larger branches fluted or buttressed, not round or symmetrical in cross section; thorns usually stout and strongly curved.

Series 9. Tenuifoliae. (C. macrosperma Ashe,etc.)

c. Boles and larger branches normally terete and symmetrical in cross section;thorns variable but not conspicuously short and curved.

d. Bark thin, pale gray, exfoliating in large plates or flakes over orange-brown or cinnamon colored inner bark. Becoming large trees in low alluvial ground; often thornless or sparingly thorny.

Series 5. Virides. (C. viridis L., etc.)

d. Bark thicker, becoming rough or more or less scaly, gray or gray-brown.

e. Bark on old boles thick and deeply ridged and fissured;branchlets rather stout, usually thorny. Becoming large trees in fertile uplands.

Series 12. Molles. (C. mollis (T. and G.)Scheele,etc.)

e. Bark on boles not deeply ridged or fissured,usually becoming more or less scaly. Small trees seldom over 8-10 m. tall.

f. Crown in well-developed trees broadly conical or depressed conical, often broader than high, lower branches often slightly depressed. Compound thorns often on boles and large branches.

g. Branchlets slender, glabrous, usually flexuous and quite thorny; lower branches dense and intricate; bark of old boles scaly, smooth and dark gray on branches.

Series 6. Crus-galli (C. crus-galli L., etc.)

g. Branchlets stoutish, the youngest slightly pubescent, armed with long slender thorns or nearly thornless. Bark on old boles brownish-gray, finely scaly.

Series 7. Punctatae (C. punctata Jacq., etc. Fig. 142).

f. Crown usually conical, rounded or irregular, rarely as broad as high.

g. Crown-irregular, open, of stoutish spreading branches; branchlets flexuous and thorny. Often fruiting as arborescent shrubs.

Series 13. Pruinosae. (C. pruinosa (Wendl.) K. Koch, etc.)

g. Crown more symmetrical, usually rounded or conical in well-developed plants. Branches more numerous and intricate.

h. Trees up to 10 m. tall with wide-spreading branches; youngest branchlets reddish brown.
Series 11. Coccineae (C. pedicellata Sarg., etc.)

h. Small trees seldom over 5-6 m. tall, with round or conical crowns; branchlets soon gray.

i. Bark close, slightly fissured and broken into small brownish-gray scales, growing usually in low moist or flat woods.
Series 10. Silvicolae. (C. iracunda Beadle, etc.)

i. Bark becoming slightly scaly, dark gray-brown, growing usually in upland thickets or rocky ground.
Series 8. Rotundifoliae. (C. chrysocarpa Ashe, etc.)

POTENTILLA L. (Rosaceae)

Small deciduous shrubs (mostly herbs). Twigs very slender, nearly round, with quickly exfoliating bark; pith small, round, brown. Buds relatively large, solitary, sessile, oblong, with about 4 striate scales. Leaf-scars alternate, very small, round, on the end of the clasping 3-nerved leaf-base; bundle-trace 1; stipules persistent.

1. P. fruticosa L. Shrubby Cinquefoil. Low bushy shrub 0.2-1 m. high, the outer bark pale, shreddy. Open ground, Labrador to Alaska, south to Pennsylvania, Iowa, South Dakota, Arizona and California.

RUBUS L.* (Rosaceae)

Rather soft-wooded usually deciduous shrubs, mostly armed with prickles, erect, trailing, or scrambling over supports. Shoots moderate, often angled, pith relatively large, round or angled, continuous. Buds moderate, sessile, oblong-ovoid. Leaf-scars

* Based on data furnished by H. A. Davis

alternate, torn and irregularly shriveled on the persistent base of the petiole; bundle-trace not discernible; stipules often persistent at the top of the petiole remnant.

Like Crataegus, this is a most difficult group taxonomically. It is believed that the few, ancient, stable species inhabiting eastern North America at the time of its settlement have given rise through crossing to numerous incipient "species", some fertile and some apomictic, many of which are spreading rapidly. An astonishing number of such "species" seem to appear in certain small areas, for example, in New England, and in the Alleghenies of West Virginia. It is impossible to describe a genus such as this except in a tentative manner, and it will be recognized that identification in winter is virtually impossible. Except for a few clear-cut species of raspberries, the following key is designed to separate the larger groups (subgenera, or sections) only.

- a. Plant habitually unarmed — 1. R. odoratus
- a. Plants habitually armed, but with thornless or nearly thornless mutations
  - b. Canes upright
    - c. Canes habitually weakly armed (Raspberries)
      - d. Plant red-hairy all over — 2. R. phoenicolasius
      - d. Plants not red-hairy
        - e. Stems bristly, not glaucous — 3. R. strigosus
        - e. Stems prickly, very glaucous — 4. R. occidentalis
    - c. Canes habitually strongly armed (sometimes nearly thornless)
      - d. Canes hispid or setose, prickles few (Bristleberries: R. *setosus* Bigel., etc.)
      - d. Canes not hispid or setose, prickles usually numerous (Blackberries)
        - e. Canes nearly without prickles or with only a few straight ones (Smooth Blackberries: *R. canadensis* L., etc.)

e. Prickles abundant, hooked or bent or at least broad-based

f. Panicle-vestiges glandular (Copsy Highbush Blackberries: R. alleghenienisis Porter, Fig. 148, etc.)

f. Panicle-vestiges not glandular (Field Highbush Blackberries: R. argutus Link, etc.)

b. Canes trailing on the ground

c. Canes hispid or bristly (Groundberries: R. hispidus L., Fig. 148, etc.)

c. Canes prickly rather than exclusively bristly (Dewberries: R. flagellaris Willd., etc.)

1. R. odoratus L. Purple Flowering Raspberry. Stems shrubby, unarmed but more or less bristly, 1-2 m. high; bark shredding. Rocky places, Quebec and Ontario, south in the mountains to Georgia and Tennessee (Fig. 143).

2. R. phoenicolasius Maxim. Wineberry. Stems biennial, long and curving, rooting at the tips, beset with long red-brown glandular hairs and weak, nearly straight prickles. Introduced from Asia, originally cultivated, now extensively naturalized (Fig. 144).

3. R. strigosus Michx. Wild Red Raspberry. (R. idaeus L. var. strigosus (Michx.) Maxim.). Stems shrubby, 1-2 m. high, densely clothed with weak glandular bristles, or the older stems with small hooked prickles. Thickets, Labrador to British Columbia, south to West Virginia, Wisconsin, Nebraska, and Wyoming (Fig. 145).

4. R. occidentalis L. Black Raspberry. Blackcaps. Stems very glaucous, recurved, often rooting at the tip, sometimes 3-4 m. long, sparingly armed with small hooked prickles. Rich thickets, Quebec to Minnesota, south to Georgia and Colorado (Fig. 146).

ROSA L. (Rosaceae)

Deciduous shrubs, mostly prickly, erect, climbing or scrambling. Shoots moderate, rounded; pith relatively large, brown,

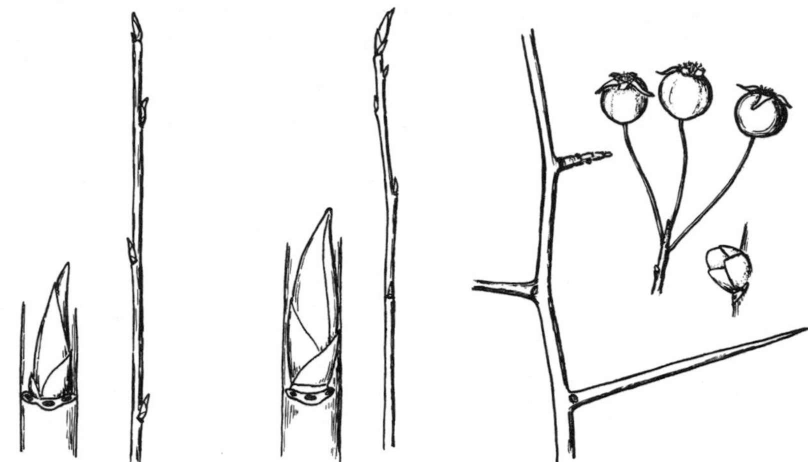

Fig. 140. Amelanchier laevis

Fig. 142. Crataegus punctata

Fig. 141. Amelanchier bartramiana

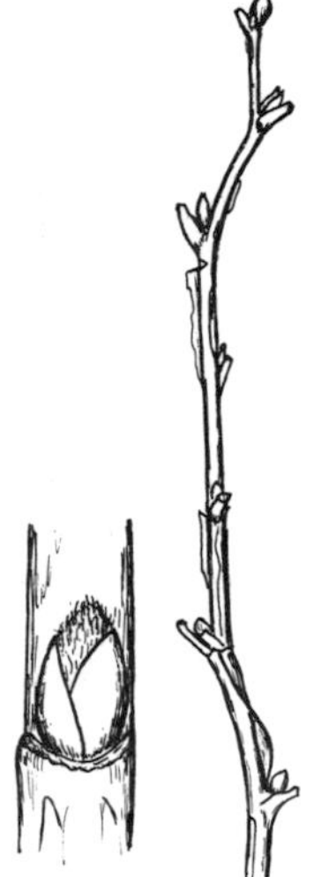

Fig. 143. Rubus odoratus

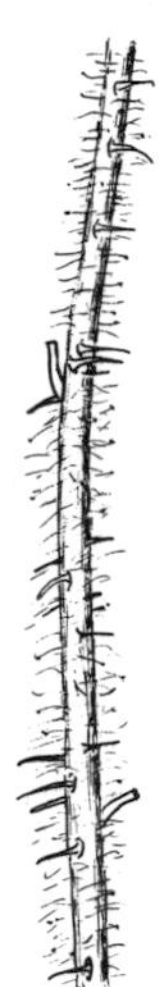

Fig. 144. Rubus phoenicolasius

Fig. 145. Rubus strigosus

rounded. Buds small, solitary, sessile, ovoid, with 3 or 4 visible scales. Leaf-scars alternate, narrow, straight or slightly curved; bundle-traces 3; stipule-scars none. Fruit a berry-like fleshy receptacle (hip) enclosing numerous achenes; the sepals, crowning the summit, are quickly deciduous or persistent into winter.

Members of this genus are so variable, as a result of extensive cultivation, that the following key should be regarded as highly tentative.

a. Stems climbing, leaning or trailing

- b. Plants of dry clearings or roadsides
  - c. Native species — 1. R. setigera
  - c. Introduced cultivated plant — 2. R. multiflora
- b. Plant of swamp thickets — 6. R. palustris

a. Stems bushy, erect

- b. Canes densely bristly and acicular-prickly — 8. R. acicularis
- b. Canes usually prickly but not densely bristly
  - c. Stems short, not averaging over 0.5 m. tall
    - d. Prickles straight, slender — 7. R. carolina
    - d. Prickles thick, more or less curved — 5. R. virginiania
  - c. Stems taller, 1-2 m. high or more
    - d. Sepals deciduous from the fruit — 4. R. canina
    - d. Sepals tardily deciduous, persisting into the winter — 3. R. eglanteria

1. R. setigera Michx. Prairie Rose. Stems more or less climbing, to 5 m. long, not bristly but armed with thick, nearly straight, scattered prickles; fruit globose, 8-10 mm. in diameter, glandular, the sepals deciduous. Thickets, Florida to Texas, north to New York, Indiana, and Kansas (Fig.149).

2. R. multiflora Thunb. Multiflora Rose. Rambler Rose. Trailing or arching, with long reclining or climbing branches;

Fig. 146. Rubus occidentalis　Fig. 147. Rubus hispidus　Fig. 148. Rubus allegheniensis

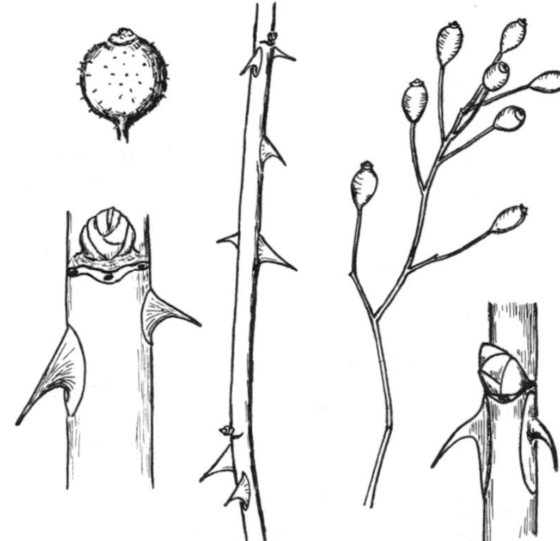

Fig. 149. Rosa setigera　Fig. 150. Rosa multiflora　Fig. 151. Rosa eglanteria

prickles hooked. Introduced from Asia, escaped from cultivation and naturalized (Fig. 150).

3. R. eglanteria L. Sweetbrier. Eglantine. Stems slender, 1-2 m. high, or forming longer wands, armed with strong hooked prickles, sometimes with scattered smaller ones; sepals tardily deciduous from the scarlet or orange hip. Introduced from Europe, escaped from cultivation and naturalized (Fig. 151).

4. R. canina L. Dog Rose. Stem 3 m. high or less, coarse, with large recurved prickles; sepals promptly falling from the ellipsoid scarlet hip. Introduced from Europe, escaped from cultivation and naturalized (Fig. 152).

5. R. virginiana Mill. Pasture Rose. Stems erect, often tall and thick, 2-20 dm. high, with large usually curved prickles; sepals soon deciduous from the red fruit. Thickets, Newfoundland to Ontario, south to Alabama and Missouri (Fig. 153).

6. R. palustris Marsh. Swamp Rose (R. carolina L.). Stems usually tall, climbing or scrambling, 3-25 dm. high, with thick straight or curved prickles; fruit globose or depressed-globose, about 8 mm. high, glandular-hispid; sepals deciduous from the depressed-globose or ellipsoid fruit. Swamps and wet thickets, New Brunswick to Ontario and Minnesota, south to Florida and Arkansas (Fig. 154).

7. R. carolina L. Low Pasture Rose. (R. serrulata of authors, not Raf.). Stems erect, 3-9 dm. high; prickles needle-like, straight; sepals deciduous from the fruit. Dry soil, Florida to Texas, north to Nova Scotia, Minnesota, and Nebraska (Fig. 155).

8. R. acicularis Lindl. Northern Rose. Canes 3-12 dm. high, densely bristly and acicular-prickly; sepals persisting and erect in fruit. Rocky slopes, Quebec to Yukon, south to Colorado, South Dakota, Michigan, and West Virginia.

PRUNUS L. (Rosaceae)

Deciduous shrubs and trees. Twigs slender or moderate, subterete or somewhat angled at the nodes, sometimes sharp-pointed; pith round or angled, pale or brown, continuous. Buds solitary or multiple, sessile, subglobose or ovoid, with about 6 visible scales. Leaf-scars alternate, half-round or elliptical, small; bundle-traces 3, usually minute; stipule-scars or vestiges present.

Fig. 152. Rosa canina

Fig. 153. Rosa virginiana

Fig. 154. Rosa palustris

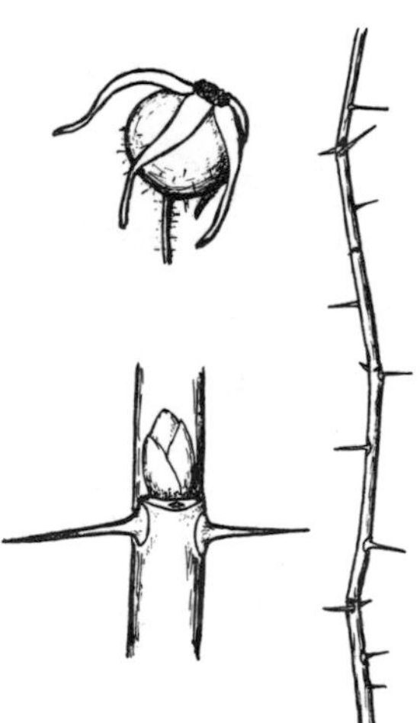

Fig. 155. Rosa carolina

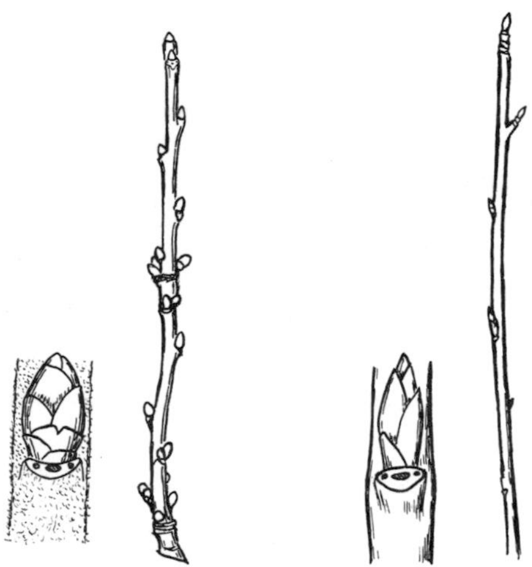

Fig. 156. Prunus maritima

Fig. 157. Prunus alleghaniensis

a. Terminal bud present
- b. Twigs red or green; buds hairy — 6. P. persica
- b. Twigs brown or gray
  - c. Twigs more or less velvety — 10. P. mahaleb
  - c. Twigs glabrous
    - d. Buds dull brown, ovoid; scales rough — 12. P. virginiana
    - d. Buds clear brown or glossy
      - e. Buds small, 2-4 mm. long
        - f. Buds 4 mm. long; scales chestnut-brown, keeled — 11. P. serotina
        - f. Buds 2 mm. long; scales red-brown, ciliate — 7. P. pensylvanica
      - e. Buds larger, 5-7 mm. long
        - f. Buds ovoid-fusiform, glossy — 8. P. avium
        - f. Buds round-ovoid, duller or darker — 9. P. cerasus

a. Terminal bud deciduous
- b. Buds about as long as thick — 5. P. angustifolia
- b. Buds elongated
  - c. Twigs velvety, becoming glabrate — 1. P. maritima
  - c. Twigs glabrous
    - d. Armed with spine-tipped twigs
      - e. Buds 3-6 mm. long; small tree — 3. P. americana
      - e. Buds 2 mm. long; usually a low straggling shrub — 2. P. alleghaniensis
    - d. Unarmed
      - e. Buds large, sometimes 10-12 mm. long; tree — 4. P. hortulana
      - e. Buds small, 2 mm. long; shrub — 2. P. alleghaniensis

1. P. maritima Marsh. Beach Plum. Low and straggling or ascending shrub 0.3-2.5 m. high, densely branched; twigs at

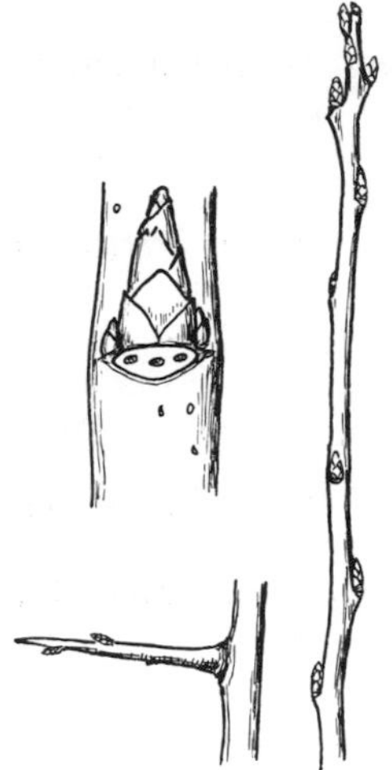

Fig. 158. Prunus americana

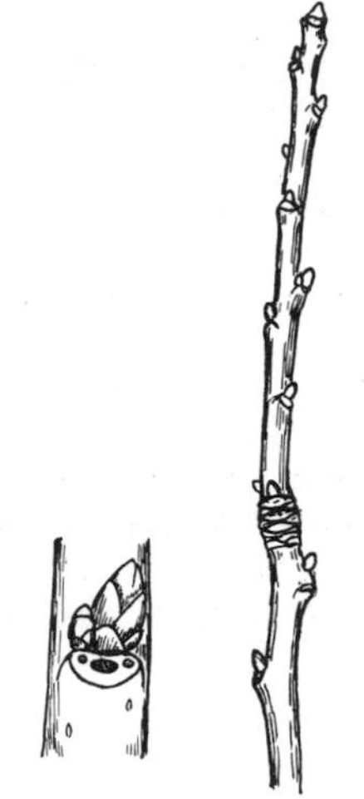

Fig. 159. Prunus hortulana

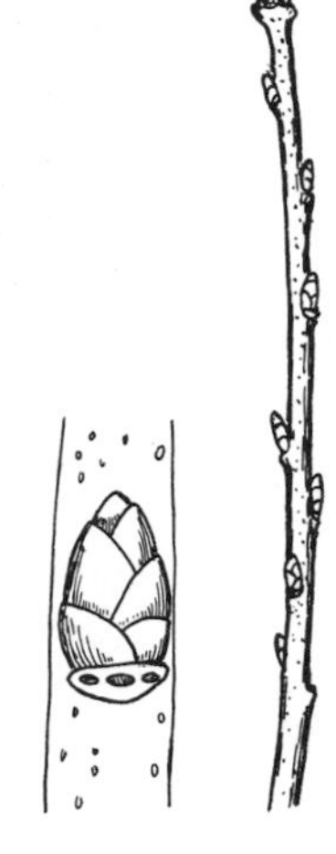

Fig. 160. Prunus angustifolia

Fig. 161. Prunus persica

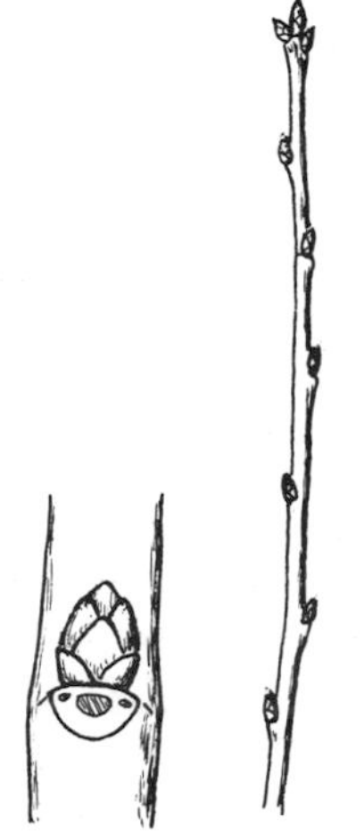

Fig. 162. Prunus pensylvanica

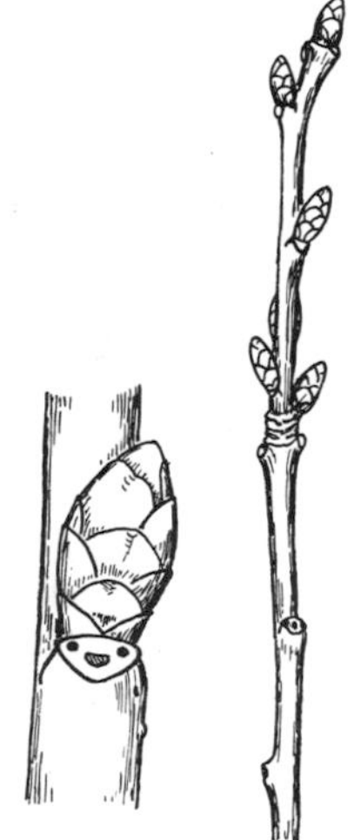

Fig. 163. Prunus avium

first pubescent, becoming glabrate; buds acute, velvety. Sandy soil, mostly in the coastal plain, Maine to Delaware (Fig. 156).

2. P. allegheniensis Porter. Alleghany Plum. A straggling shrub 1-5 m. high, sometimes armed with sharp-pointed twigs; bark brown; twigs reddish-brown. Dry soil, mostly in the mountains, Connecticut to Virginia and West Virginia (Fig. 157).

3. P. americana Marsh. Wild Plum. A shrub or tree 3-10 m. high, bark dark, shaggy; twigs orange-brown, sharp-pointed, glabrous; buds acute, 3-6 mm. long, chestnut-brown. Thickets, Florida to New Mexico, north to New England, Ontario, Manitoba, Wyoming, and Utah; also in Mexico (Fig. 158).

4. P. hortulana Bailey. Wild Goose Plum. A small unarmed tree, 10 m. high or less; twigs glabrous, reddish-brown; buds obtuse, chestnut-brown. Bottomlands, Indiana to Iowa, south to Alabama and Oklahoma (Fig. 159).

5. P. angustifolia Marsh. Chickasaw Plum. A shrub 2-5 m. high, not very thorny; buds half covered by the ciliate leaf-cushion; twigs slender, red, lustrous. Dry thickets, Florida to Texas, north to New Jersey and Missouri (Fig. 160).

6. P. persica (L.) Batsch. Peach. A small tree, up to 8 m. tall, the reddish or green twigs glabrous; buds hairy. Introduced from Asia, much cultivated, and frequently escaped (Fig. 161).

7. P. pensylvanica L.f. Fire Cherry. Pin Cherry. Bird Cherry. A tree 6-20 m. high, with light red-brown bark; twigs glabrous, slender, reddish, shining. Dry woods, recent burns and clearings, Labrador to British Columbia, south to Colorado, South Dakota, Iowa, and in the mountains to North Carolina (Fig. 162).

8. P. avium L. Sweet Cherry. A large tree, sometimes 20 m. high, with pyramidal crown and reddish-brown bark; twigs rather thick, buds large, glossy, ovoid-fusiform. Introduced from Eurasia, often spread from cultivation and naturalized (Fig. 163).

9. P. cerasus L. Sour Cherry. Tree to 24 m. high, of pyramidal habit; branchlets rather thick; buds large, dull, round-ovoid. Introduced from Asia, spread from cultivation and naturalized (Fig. 164).

10. P. mahaleb L. Mahaleb. Perfumed Cherry. Small tree to 10 m. high, with open branching; bark aromatic; twigs tomen-

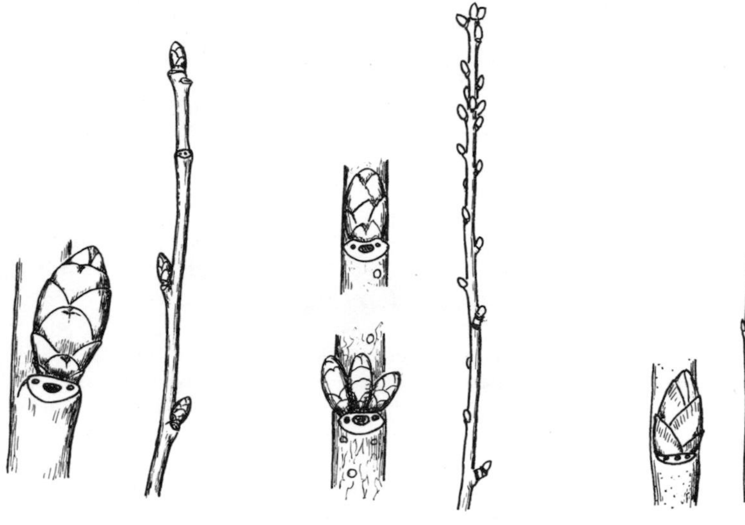

Fig. 164. Prunus cerasus

Fig. 165. Prunus mahaleb

Fig. 166. Prunus serotina

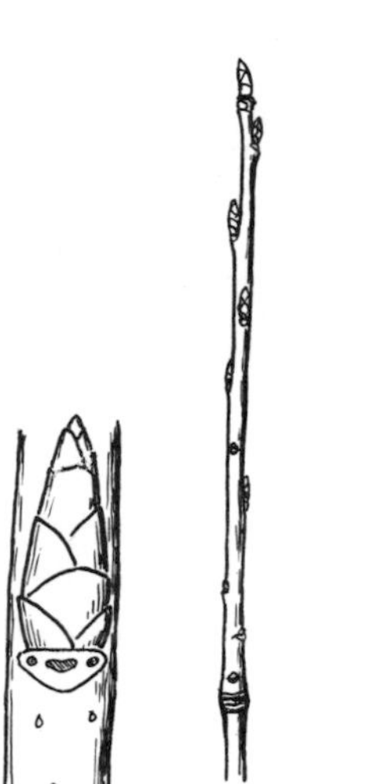

Fig. 167. Prunus virginiana

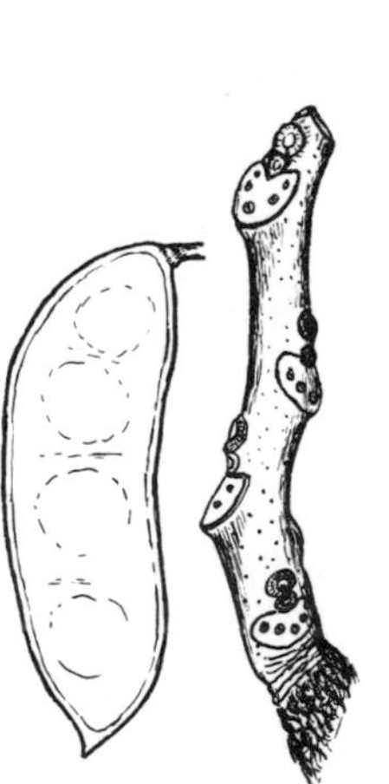

Fig. 168. Gymnocladus dioica

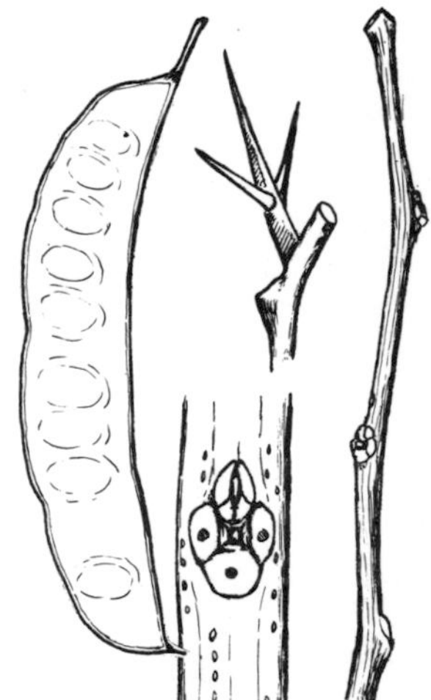

Fig. 169. Gleditsia triacanthos

tulose; buds round-ovoid, spreading. Introduced from Eurasia, spread from cultivation and naturalized (Fig. 165).

11. P. serotina Ehrh. Wild Black Cherry. A tree 20-35 m. tall, the trunk 10-15 dm. in diameter; bark on old trees very rough and black, on the branches smoother and gray-brown; inner bark strongly and unpleasantly scented; twigs reddish-brown, buds about 4 mm. long. Woods and open fields, Florida to Texas and Mexico, north to New Brunswick, Minnesota and North Dakota (Fig. 166).

12. P. virginiana L. Choke Cherry. A tall shrub or small tree, 15 m. high, with grayish bark, the inner layers with a rank disagreeable odor; twigs glabrous. Thickets, Newfoundland to Saskatchewan, south to North Carolina, Missouri and Kansas (Fig. 167).

GYMNOCLADUS Lam. (Leguminosae)

Large, deciduous, rough-barked tree. Twigs thick, terete or 3-sided; pith large, round, continuous, salmon-colored. Buds superposed, in raised silky pits; end-bud lacking. Leaf-scars alternate, large, heart-shaped; bundle-traces 3 or 5, rather indefinite and divided; stipule-scars minute or lacking.

1. G. dioica (L.) K. Koch. Kentucky Coffeetree. A tree 35 m. tall, or less; bark varying from gray to dark brown, characterized by sharp horny ridges; twigs very thick, finely pubescent, often coated with a crusty film, marked with fine lenticels; buds small, downy, deeply sunken, surrounded by an incurved hairy ring of bark; pods 1-2 dm. long, 3-4 cm. broad, often persisting until late winter. Rich woods, New York to South Dakota, south to Tennessee and Missouri; cultivated and naturalized eastwards and northwards (Fig. 168).

GLEDITSIA L. (Leguminosae)

Spreading deciduous trees usually armed with large branched spines arising above the axils and persisting on the trunk. Twigs zigzag, roundish; pith rounded, continuous, pale or light pink. Buds glabrous, sessile, superposed, more or less covered by the torn margin of the leaf-scar; end-bud lacking. Leaf-scars alternate, relatively large, shield-shaped; bundle-traces 3; stipule-scars none. Fruit a large legume (to 4 dm. long).

1. G. triacanthos L. Honey-Locust. A tree with a maximum height of 50 m., usually much smaller; bark on young trunks smooth

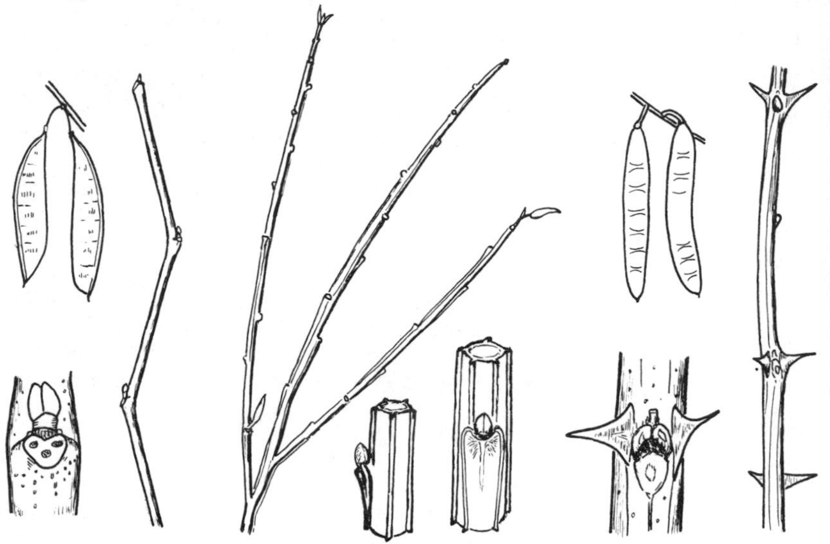

Fig. 170. Cercis canadensis

Fig. 171. Cytisus scoparius

Fig. 172. Robinia pseudo-acacia

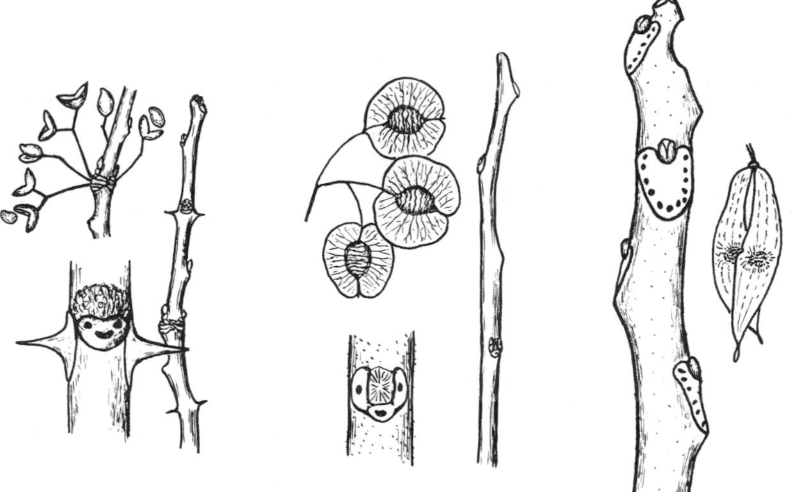

Fig. 173. Zanthoxylum americanum

Fig. 175. Ailanthus altissima

Fig. 174. Ptelea trifoliata

and marked with large oblong conspicuous lenticels, on older trunks roughened by scattered shallow fissures and thick firm broad plates; trunk thickly beset with large branching thorns (thorns absent in f. inermis (Pursh) Schneid.). Rich woods, New York to South Dakota, south to Florida and Texas (Fig. 169).

CERCIS L. (Leguminosae)

Deciduous shrubs or small trees. Twigs moderate, zigzag, subterete; pith roundish, continuous, pale or pinkish. Buds glabrous, the flower-buds superposed and stalked, the leaf-buds sessile, ovoid or obovoid, with 2 scales (or several in the flower buds); end-bud lacking. Leaf-scars alternate, 2-ranked, somewhat raised, fringed at the top, obtusely triangular, with 3 small decurrent ridges; bundle-traces 3; stipule-scars lacking.

1. C. canadensis L. Redbud. Shrub or small tree 8-10 m. high; bark thin, with shallow grooves and dark brown scaly ridges; twigs dark brown or black; pith round, pale or pinkish, continuous; buds 2-3 mm. long; pods brown, 6-8 cm. long, 12 mm. wide, persistent in winter. Rich woods, Florida to Texas and Mexico, north to Connecticut, Ontario and Wisconsin (Fig. 170).

CLADRASTIS Raf. (Leguminosae)

Handsome trees with yellow wood, smooth bark, and deciduous leaves. Twigs moderate, terete, somewhat zigzag; pith moderate, round, continuous, pale. Buds sessile, either solitary and scaly or in superposed groups and not distinctly scaly; end-bud lacking. Leaf-scars alternate, 2-ranked, half-round or horseshoe-shaped and encircling the bud; bundle-traces 3 or 5; stipule-scars lacking.

1. C. lutea (Michx. f.) K. Koch. Yellowwood. Virgilia. A small tree to 17 m. high; twigs red-brown; buds short, scarcely 5 mm. long. Rich woods, especially on limestone outcrops, Missouri, Illinois and Kentucky, south to Georgia and Alabama.

CYTISUS L. (Leguminosae)

Small shrubs, commonly deciduous. Twigs slender, terete or more usually ribbed or grooved; pith small, roundish, continuous. Buds small, solitary, sessile, round-ovoid, with about 4 scales. Leaf-scars alternate, elevated on a leaf-cushion, minute; bundle-trace 1, indistinct; stipules or minute stipule-scars present.

1. C. scoparius (L.) Link. Scotch Broom. A shrub 1-2 m. high, with glabrous stiff green branches; twigs finely granular, almost winged on the ridges. Introduced from Europe and naturalized (Fig. 171).

ROBINIA L. (Leguminosae)

Deciduous trees or shrubs. Twigs often prickly, more or less angled, zigzag; pith round, continuous. Buds small, superposed, covered by the leaf-scar; terminal bud lacking. Leaf-scars alternate, broadly triangular, consisting of a membrane that later splits open, revealing the buds; bundle-traces 3; stipules modified as bristles or prickles which enlarge and persist for several years.

a. Twigs glabrous or nearly so 1. R. pseudo-acacia
a. Twigs bristly, glandular or viscid 2. R. viscosa

1. R. pseudo-acacia L. Black Locust. A tree to 25 m. high, but generally smaller; bark rough, deeply furrowed, dark brown; twigs moderate, somewhat zigzag, greenish to reddish-brown, usually with two stipular spines at each node; pods dark brown, 5-10 cm. long, 10-12 mm. wide, often remaining on trees through the winter. Woods and thickets, Georgia to Louisiana, north to Pennsylvania, Indiana, and Oklahoma, now widely introduced and naturalized (Fig. 172).

2. R. viscosa Vent. Clammy Locust. A small tree, to 12 m. high, with dark red-brown glandular-viscid twigs; stipular spines small or lacking; pods clammy. Dry woods, in the mountains, North Carolina to Georgia and Alabama; cultivated northwards.

WISTERIA Nutt. (Leguminosae)

Twining deciduous shrubs. Stems moderate, somewhat fluted; pith moderate, white or brown, round, continuous. Buds moderate, sessile, oblong, acute, nearly surrounded by the outer scale. Leaf-scars alternate, elliptical, raised, with a protuberance at each side; bundle-trace 1; stipule-scars lacking.

1. W. frutescens (L.) Poir. Wisteria. Stems up to 12 m. long; branchlets glabrous or nearly so. Banks of streams and swamps, on the coastal plain, Virginia to Florida and Alabama.

ZANTHOXYLUM Gmel. (Rutaceae)

Deciduous aromatic shrubs or small tree, armed with prickles often paired at the nodes (as in Robinia) but not believed to be modified stipules. Twigs moderate, rounded; pith continuous, creamy-white. Buds moderate, superposed, sessile, globose, woolly. Leaf-scars alternate, broadly triangular or 3-lobed; bundle-traces 3; stipule-scars none.

1. Z. americanum Mill. Toothache-Tree. A shrub or small tree, up to 8 m. high; nodal prickles widened at the base; leaf-scars with a conspicuous articular membrane; buds red-rusty. Rich woods, Quebec to North Dakota, south to Georgia and Oklahoma (Fig. 173).

PTELEA L. (Rutaceae)

Deciduous shrubs or small trees. Twigs moderate, warty-dotted; pith large, continuous, white. Buds moderate, superposed, silvery-silky, low-conical, sessile, breaking through the leaf-scars; terminal bud lacking. Leaf-scars alternate, relatively large, U-shaped when torn by the buds; bundle-traces 3; stipule-scars none.

1. P. trifoliata L. Hoptree. A shrub or small tree, up to 8 m. high, without prickles; twigs buff, glabrous (or pubescent in the var. mollis T. and G.); fruit a yellowish suborbicular samara with broad veiny wings, 1.8-2.5 cm. in diameter, persistent in winter. Thickets, Quebec to Ontario and Nebraska, south to Florida, Texas, Arizona, and Mexico (Fig. 174).

AILANTHUS Desf. (Simaroubaceae)

Smooth-barked loosely-branched deciduous trees. Twigs very thick, somewhat 3-sided, with prominent lenticels; pith large, roundish. Buds solitary, sessile, relatively small; terminal bud lacking. Leaf-scars alternate, shield-shaped, large; bundle-traces 9 or more; stipule-scars none.

1. A. altissima (Mill.) Swingle. Tree of Heaven. (A. glandulosa Desf.). A tree to 20 m. high, having relatively smooth bark with pale stripes; twigs puberulent. Introduced from Asia, widely naturalized from cultivation, often weed-like and competing with native trees (Fig. 175).

RHUS L. (Anacardiaceae)

Deciduous shrubs or small trees, or climbing by aerial roots, with milky, sometimes poisonous sap. Twigs round or obscurely 3-sided, slender to quite thick; pith large. Buds moderate or small, solitary, sessile, round-ovoid; terminal bud often lacking. Leaf-

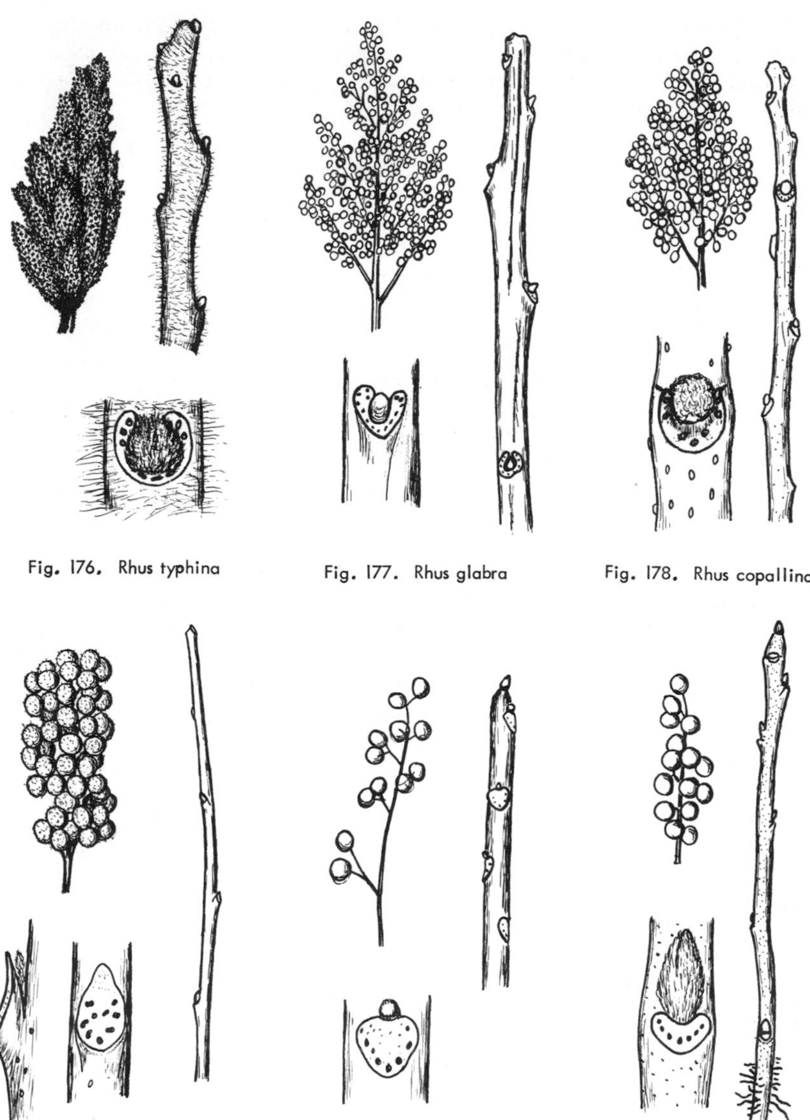

Fig. 176. Rhus typhina

Fig. 177. Rhus glabra

Fig. 178. Rhus copallina

Fig. 179. Rhus aromatica

Fig. 180. Rhus vernix

Fig. 181. Rhus radicans

scars round to crescent-shaped, or horseshoe-shaped and encircling the bud; bundle-traces numerous in the round leaf-scars, but often 3, 5, or 9 single traces or groups of traces in the narrow leaf-scars; stipule-scars none.

a. Leaf-scars horseshoe-shaped or U-shaped

  b. Leaf-scars horseshoe-shaped, nearly encircling the bud; twigs thick

    c. Twigs velvety-hairy, rounded. . . . 1. R. typhina
    c. Twigs glabrous, somewhat 3-sided . . . 2. R. glabra

  b. Leaf-scars U-shaped; twigs terete, puberulent . . . 3. R. copallina

a. Leaf-scars round, or broadly crescent or shield-shaped

  b. Leaf-scars circular, distinctly raised, covering the small yellow hairy buds; twigs slender . . . 4. R. aromatica
  b. Leaf-scars broadly crescent or shield-shaped

    c. Erect; twigs thick; buds sessile . . . 5. R. vernix
    c. Climbing or bushy-spreading; twigs slender; buds stalked . . . 6. R. radicans

1. *R. typhina* L. *Staghorn Sumac.* A shrub or small tree, 10 m. high or less; twigs densely hairy, the hairs concealing the lenticels; fruits in dense clusters, globular, covered with red hairs, persistent in winter. Thickets, Quebec to Minnesota, south to North Carolina and Iowa (Fig. 176).

2. *R. glabra* L. *Smooth Sumac.* A shrub 0.5–8 m. high, with smooth, glaucous, somewhat 3-sided twigs; fruits in dense clusters, covered with reddish viscid hairs, persistent in winter. Dry soil, Maine and Quebec to British Columbia, south to Florida and Arizona (Fig. 177).

3. *R. copallina* L. *Dwarf Sumac.* Shrub or small tree, 10 m. high or less; twigs reddish-brown, downy-puberulent; fruits covered with crimson hairs, persistent in winter. Dry soil, Maine

to Illinois, south to Florida and Texas (Fig. 178).

4. R. aromatica Ait. Aromatic Sumac (R. canadensis Marsh.). A straggling fragrant shrub 1-2 m. high, with ascending or diffuse branches; twigs pubescent; buds small, yellow, hairy, covered by the leaf-scars; fruits globose, covered with long soft red hairs. Dry soil, Quebec to Illinois and Nebraska, south to Florida and Texas (Fig. 179).

5. R. vernix L. Poison Sumac. (Toxicodendron vernix Ktze.). Shrub or small tree up to 8 m. high, with smooth, slightly streaked, light to dark gray bark; terminal bud present, glabrate, moderate, 5 mm. or less long; fruit white, conspicuous all winter. Twigs poisonous to the touch, causing a dermatitis. Swamps, Quebec to Minnesota, south to Florida and Texas (Fig. 180).

6. R. radicans L. Poison-Ivy. (R. toxicodendron of authors, not L.; Toxicodendron radicans Ktze.). A shrub, erect and bushy, scrambling over rocks, or climbing by aerial roots; twigs sparingly pubescent or glabrate; fruits greenish-white. Poisonous to the touch, in winter as well as summer. Thickets and fence-rows, Quebec to British Columbia, south to Florida, Texas, New Mexico and Arizona; often too abundant (Fig. 181).

ILEX L. (Aquifoliaceae)

Evergreen or deciduous shrubs or small trees. Twigs slender, often developed as spurs with densely crowded leaf-scars; pith small, continuous. Buds small, usually superposed, sessile. Leaf-scars alternate, crescent-shaped; bundle-traces 1; stipule-scars minute. Fruit fleshy, red (or sometimes yellow), composed of 4 to 6 hard nutlets which fit together somewhat like the sections of an orange; this might be regarded as a several-seeded drupe (see p. 11).

a. Leaves evergreen

b. Leaves 0.4-1 cm. long ... 1. I. opaca
b. Leaves 1.5-5 cm. long ... 5. I. glabra

a. Leaves deciduous

b. Buds more or less appressed

c. Buds glabrous except at apex; fruits on short stalks ... 3. I. montana
c. Buds pubescent; fruits on long stalks ... 2. I. collina

b. Buds not appressed 4. I. verticillata

1. I. opaca Ait. American Holly. Small tree, 6-20 m. high, bark close, rough, gray, with inconspicuous lenticels; twigs slender; buds short, blunt, downy; leaves evergreen, elliptical, thick, 5-10 cm. long, smooth, with wavy margin and remote spiny teeth or occasionally entire; fruit red or sometimes yellow, 7-10 mm. in diameter. Moist woods, Florida to Texas, north to Massachusetts, Pennsylvania, Missouri and Oklahoma (Fig. 182).

2. I. collina Alexander. Long-stalked Holly. Shrub or small tree 3-4 m. tall; branches spreading, bark smooth, gray; twigs gray, glabrous; buds somewhat appressed, pubescent; lenticels conspicuous; fruits red, rarely yellow, 7-10 mm. in diameter. Woods, Virginia and West Virginia (Fig. 183).

3. I. montana T. and G. Mountain Holly. Shrub or small tree, up to 8 m. high; bark thin, rough and warty, brownish-gray, with numerous lenticels; twigs smooth, reddish-brown, becoming gray, enlarged at the nodes, with decurrent ridges running down from the leaf-scars; buds appressed, pointed, 2 mm. long, the scales ovate, keeled, sharp-pointed, light-brown, finely hairy at the apex; fruits red, about 1 cm. in diameter. Moist woods, in the mountains, New York to Georgia and Tennessee; also in Japan (Fig. 184).

4. I. verticillata (L.) Gray. Black-Alder. Winterberry. Whorled Holly. Shrub or small tree 0.5-6 m. high; bark smooth, ashen; twigs slender, smooth or slightly pubescent; buds spreading, blunt, the scales obtuse; fruits red, rarely yellow, 5-7 mm. in diameter, so crowded as to appear whorled. Moist soil, Newfoundland to Minnesota, south to Georgia and Missouri (Fig. 185).

5. I. glabra (L.) Gray. Inkberry. Gallberry. A shrub to 3 m. high; branchlets ashy-puberulent; leaves evergreen, coriaceous, lustrous, lanceolate to oblong, mostly blunt, crenate or crenate-serrate, 1.5-5 cm. long, 0.7-2 cm. wide; drupes black, globose, 6 mm. in diameter. Sandy soil, in the coastal plain, Louisiana to Florida, north to Nova Scotia (Fig. 186).

NEMOPANTHUS Raf. (Aquifoliaceae)

Deciduous shrub with light-colored bark. Twigs slender, often remaining rather short; pith small, continuous. Buds ovoid, solitary, small, sessile, pointed at the apex, with 2 ciliate scales. Leaf-scars raised, triangular or crescent-shaped; bundle-trace 1; stipule-scars none. Fruit red, long-stalked, persistent in winter.

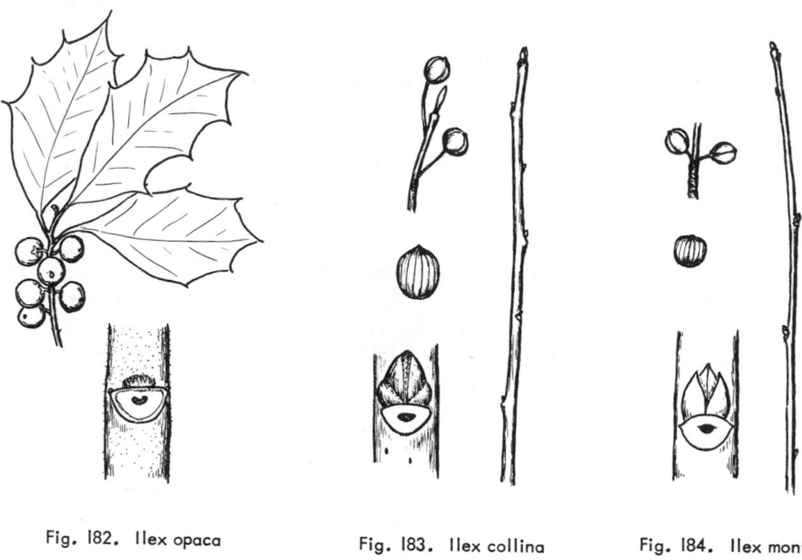

Fig. 182. Ilex opaca

Fig. 183. Ilex collina

Fig. 184. Ilex montana

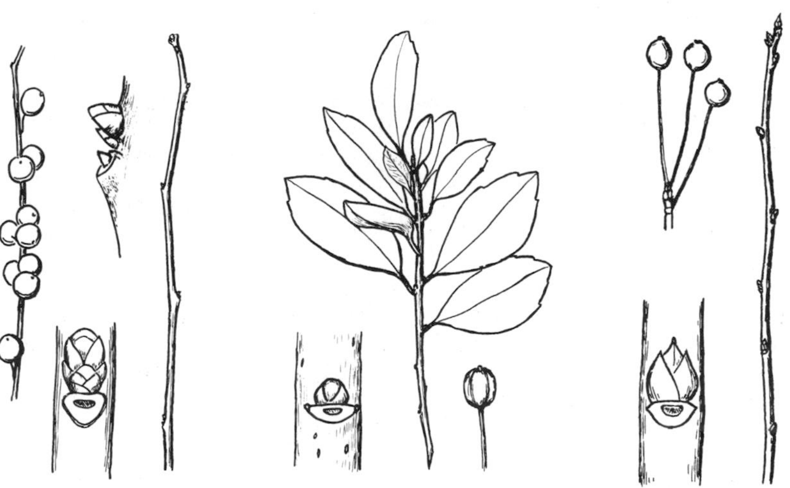

Fig. 185. Ilex verticillata

Fig. 186. Ilex glabra

Fig. 187. Nemopanthus mucronata

1. N. mucronata (L.) Trel. Wild Holly. Erect shrub, 0.3-3 m. high, with ash-colored bark; twigs slender, glaucous-purplish, becoming gray; pith small, continuous; fruits red or pale yellow. Swamps, Newfoundland to Minnesota, south to Illinois, Ohio and West Virginia (Fig. 187).

EUONYMUS L. (Celastraceae)

Shrubs or small trees, sometimes creeping, mostly deciduous. Branchlets 4-sided or 4-lined below the nodes, usually green; pith rounded or angled, greenish, porous or becoming excavated in older stems. Buds small to fairly large, solitary, sessile, with 3 to 5 pairs of scales. Leaf-scars opposite, or the members of pairs somewhat separated, half-elliptical; bundle-trace 1; stipule-scars minute, usually indistinct. Fruit a capsule which splits in autumn revealing the crimson or orange-red seeds, persistent into winter.

a. Twigs rounded, but often 4-lines — 1. E. atropurpureus
a. Twigs 4-lined

  b. Erect bushy shrub — 2. E. americanus
  b. Low and procumbent — 3. E. obovatus

1. E. atropurpureus Jacq. Burning Bush. Wahoo. Shrub up to 4 m. high, with greenish bark; buds oblong, scales about 5, oblong, loose; pods smooth. Rich woods, Ontario to Montana, south to Alabama and Oklahoma; cultivated and naturalized elsewhere (Fig. 188).

2. E. americanus L. Strawberry Bush. Low shrub, upright or straggling, 2.5 m. high, glabrous; branches green; pods warty, on a slender stalk. Rich woods, Florida to Texas, north to New York, Illinois, Missouri, and Oklahoma (Fig. 189).

3. E. obovatus Nutt. Trailing Strawberry Bush. Trailing shrub with rooting branches; upright branches 3-6 dm. high, green; buds fusiform; pods warty. Rich woods, New York to Michigan, south to Tennessee and Missouri (Fig. 190).

PACHISTIMA Raf. (Celastraceae)

Low evergreen shrubs. Branchlets very slender, somewhat 4-sided, the bark becoming transversely-checked; pith minute, round, porous, Buds solitary, sessile, ovoid, appressed, small, the lateral with about 2 pairs of scales, the terminal larger with more visible scales. Leaves small, nearly sessile. Leaf-scars opposite,

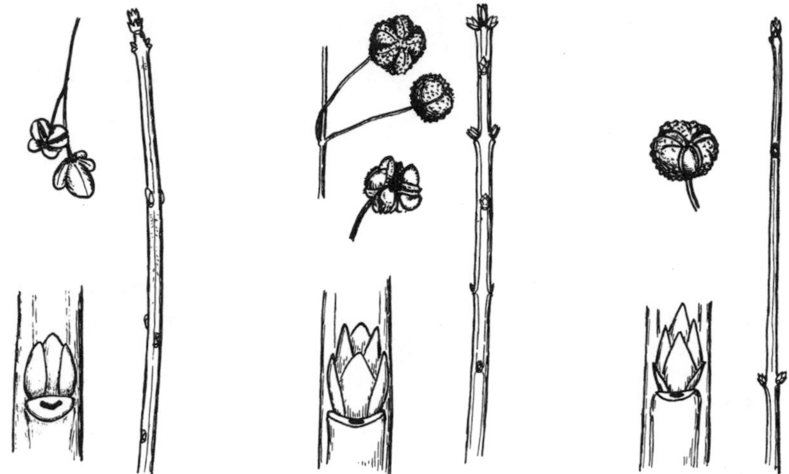

Fig. 188. Euonymus atropurpureus

Fig. 189. Euonymus americanus

Fig. 190. Euonymus obovatus

Fig. 191. Pachistima canbyi

Fig. 192. Celastrus scandens

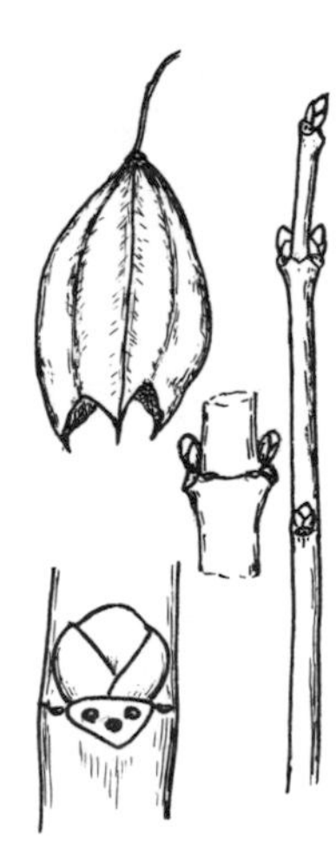

Fig. 193. Staphylea trifolia

minute, half-elliptical; bundle-trace 1, indistinct; stipule-scars none.

1. P. canbyi Gray. Canby's Mountain-lover. Dwarf shrub with rooting branches ascending to 25 cm. ; leaves linear, serrulate, obtuse, revolute, 6-25 mm. long; fruit a pod about 4 mm. long. Rocky limestone slopes, Virginia, West Virginia, Kentucky, and Ohio (Fig. 191).

CELASTRUS L. (Celastraceae)

Deciduous woody climbers. Stems terete, slender; pith large, continuous, round, white. Buds small, solitary, sessile, subglobose, with about 6 mucronate scales. Leaf-scars alternate, half-elliptical; bundle-trace 1; stipule-scars minute.

1. C. scandens L. Climbing Bittersweet. A twining shrub, climbing to 7 m. or higher; stems and buds glabrous, brownish-gray; fruit a globose pod opening in autumn and exposing the ornamental scarlet seeds, persistent into winter. Thickets, Quebec to Manitoba, south to Georgia, Louisiana, and Oklahoma (Fig. 192).

STAPHYLEA L. (Staphyleaceae)

Deciduous shrubs or small trees. Twigs moderate, rounded, glabrous; pith rather large, continuous, white. Buds solitary, sessile, ovoid, glabrous; terminal bud usually lacking. Leaf-scars opposite (sometimes somewhat separated), half-round; bundle-traces 3 or broken into 5 or 7 or more; stipule-scars rounded or elongated. Fruit an inflated capsule persistent in winter, the seeds rattling when the capsule is shaken.

1. S. trifolia L. Bladdernut. An upright shrub up to 5 m. high, with smooth striped bark; buds with 4 blunt exposed scales; pods 3-lobed, 3-4 cm. long. Rich thickets, Quebec to Minnesota, south to Georgia, Arkansas and Oklahoma (Fig. 193).

ACER L. (Aceraceae)

Deciduous shrubs or trees. Twigs moderate, rounded; pith continuous, pale. Buds moderate, solitary or collaterally multiple, ovoid or conical, sessile or stalked, with 1 or several pairs of exposed scales. Leaf-scars opposite (very exceptionally whorled in some individuals), curved or U-shaped; bundle-traces usually 3, but sometimes 5, 7, or 9, or more numerous; stipule-scars none.

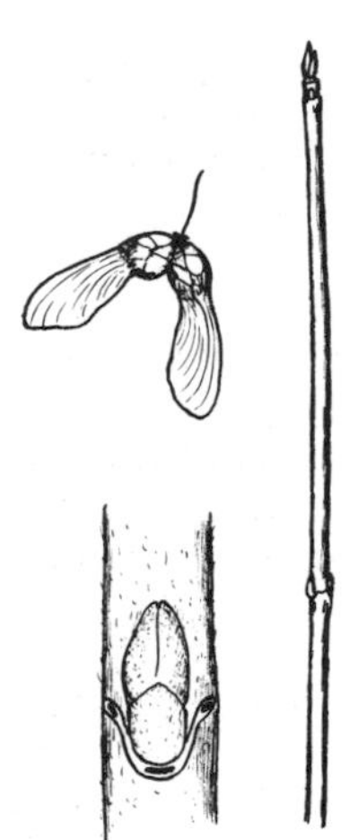

Fig. 194. Acer spicatum

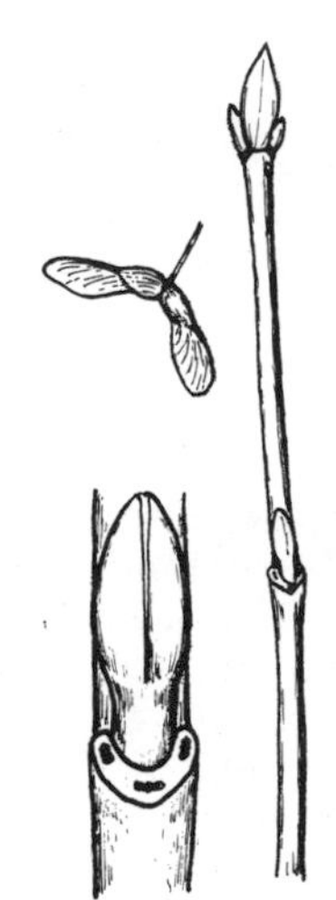

Fig. 195. Acer pensylvanicum

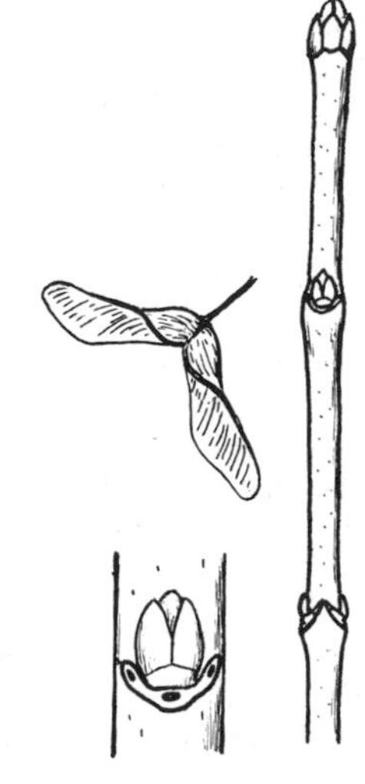

Fig. 196. Acer platanoides

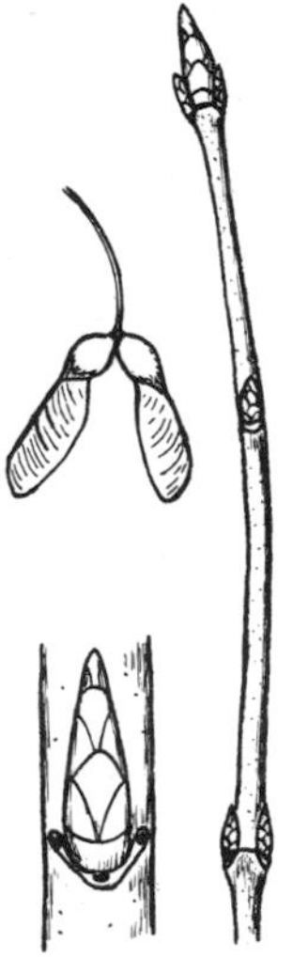

Fig. 197. Acer saccharum

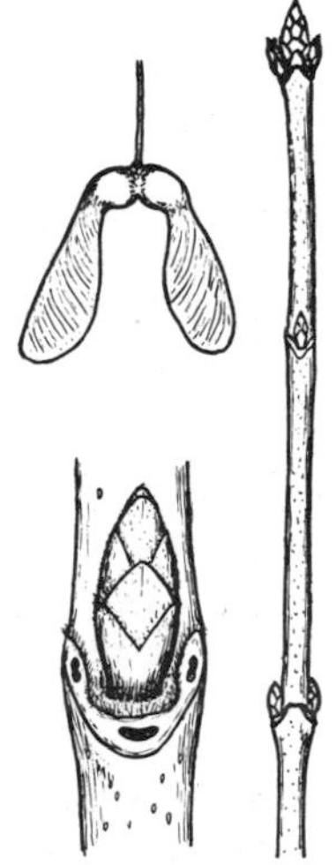

Fig. 198. Acer nigrum

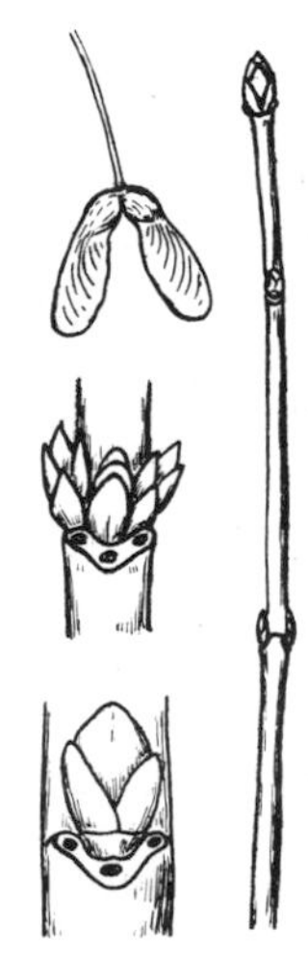

Fig. 199. Acer rubrum

a. Bud-scales 2, valvate

- b. Buds glabrous, thick, blunt, 8-10 mm. long — 1. A. pensylvanicum
- b. Buds puberulent, short, slender, 5 mm. long — 2. A. spicatum

a. Bud-scales more than 2

- b. Buds red, brown, or black; twigs not green
  - c. Terminal buds usually 5 mm. or more long; sap milky — 3. A. platanoides
  - c. Terminal buds usually less than 5 mm. long; sap clear
    - d. Buds brown or black; scales 6 or more
      - e. Twigs glossy-buff or reddish-brown; buds dark, glabrate — 4. A. saccharum
      - e. Twigs dull straw-color, smooth and shiny, with somewhat prominent lenticels; buds straw-color, hairy — 5. A. nigrum
    - d. Buds reddish or orange; scales about 4
      - e. Twigs bright chestnut-brown; inner bark with a rank odor — 7. A. saccharinum
      - e. Twigs red and lustrous; inner bark not rank — 6. A. rubrum
- b. Twigs green; buds covered with a dense white pubescence — 8. A. negundo

1. A. spicatum Lam. Mountain Maple. Shrub or small tree up to 10 m. high, with thin, rather smooth, brown or grayish-brown bark mottled with dingy gray patches; young twigs grayish-pubescent; buds short-stalked, small, about 5 mm. long; terminal buds larger. Cool woods, Newfoundland to Saskatchewan, south to Iowa, Ohio, and the mountains of Georgia and Tennessee (Fig. 194).

2. A. pensylvanicum L. Striped Maple. Moosewood. A small tree up to 10 or 12 m. high, with smooth bark, becoming conspic-

uously striped with white lines; twigs smooth, thick, green, changing to red; lenticels few; buds stalked, about 1 cm. long, tapering but blunt-pointed, red, glossy, glabrous except for the ciliate margins of the scales; leaf-scars broadly U-shaped, almost encircling the twig. Cool woods, Quebec to Manitoba, south to Michigan, Ohio, and the mountains of Tennessee and Georgia (Fig. 195).

3. A. platanoides L. Norway Maple. Tree to 30 m. high; bark rough, broken; twigs thick, with milky sap; terminal bud plump, large and conspicuous, usually more than 5 mm. long. Introduced from Europe, much planted and sometimes escaping (Fig. 196).

4. A. saccharum Marsh. Sugar Maple. Tree to 40 m. high, with gray, furrowed bark; twigs slender, dull, smooth, reddish-brown to orange-brown; buds brown, conical, sharp-pointed, glabrous or hairy at the apex, the terminal one 5 mm. or less in length; leaf-scars U-shaped or V-shaped, nearly encircling the stem. Rich woods, Quebec to Manitoba, south to Georgia, Mississippi, Arkansas, and Texas (Fig. 197).

5. A. nigrum Michx. f. (A. saccharum var. nigrum (Michx. f.) Britt.). Black Sugar Maple. Black Maple. A tree 40 m. high, with dark gray bark; twigs slender, glossy, buff; lenticels prominent; buds dark, hairy; leaf-scars U-shaped or V-shaped. Rich woods, Quebec to Minnesota and South Dakota, south to Georgia and Louisiana (Fig. 198).

6. A. rubrum L. Red Maple. Tree up to 40 m. high; bark on young trunks smooth and light gray, on older trunks dark grayish and rough; twigs rather slender, green, becoming glossy red as winter progresses; buds reddish, obtuse, accessory flower buds developing towards spring; flowers red, beginning to bloom in March. Rich soil, Newfoundland to Manitoba, south to Florida and Texas (Fig. 199).

7. A. saccharinum L. Silver Maple. A tree up to 40 m. high; bark on young trunks smooth and light gray, on older trunks brown and shallowly fissured; crushed twigs with a rank unpleasant odor; buds reddish, obtuse, sessile or short-stalked; flower buds spherical, accessory; margin of scales ciliate, often lighter in color; flowers greenish-yellow or reddish, beginning to open in February. River-banks and bottomlands, New Brunswick to Minnesota and South Dakota, south to Florida and Arkansas (Fig. 200).

8. A. negundo L. Box Elder. Tree up to 20 m. high; bark dark gray or brown, divided into broad ridges; twigs moderate, green, smooth, sometimes glaucous, with scattered, rather

prominent lenticels; buds sessile or short-stalked, large, ovoid, the terminal acute and the lateral obtuse, covered with gray hairs; leaf-scars V-shaped, encircling the stem so that the adjacent edges of opposite scars meet at a very sharp angle; bundle-traces usually 3. Riverbanks, Florida to Texas, Arizona and Nevada, north to Maine, Manitoba, Saskatchewan, and Alberta (Fig. 201).

AESCULUS L. (Hippocastanaceae)

Deciduous trees or shrubs. Twigs quite thick; pith large, continuous. Buds, especially the terminal, quite large, solitary, ovoid, sessile, with 6 or more pairs of scales. Leaf-scars opposite, large, shield-shaped or triangular; bundle-traces 3 or in 3 groups, sometimes 7 or 9; stipule-scars none. The twigs sometimes end in an inflorescence - or fruit-scar (see p. 7)

a. Buds brownish, not gummy

 b. Bark rough, soft and corky; ill-scented; fruit spiny — 1. A. glabra
 b. Bark smooth and firm, not ill-scented; fruit smooth — 2. A. octandra

a. Buds nearly black, gummy — 3. A. hippocastanum

1. A. glabra Willd. Fetid Buckeye. Ohio Buckeye. Small tree up to about 20 m. high; bark gray, breaking into plates, exhaling an unpleasant odor; terminal bud 1.5-1.8 cm. long, pointed, the outer bud scales reddish-brown, finely hairy on the margins; fruit spiny. Rich woods, Pennsylvania to Nebraska, south to Alabama and Oklahoma (Fig. 202).

2. A. octandra Marsh. Sweet Buckeye. A tree up to 25 m. high, with light brown to grayish-brown bark breaking up into numerous thin irregular scales; terminal bud 1.5 cm. long, somewhat pointed, the outer scales reddish-brown, covered with a thin bluish bloom; fruit smooth. Rich woods, Pennsylvania to Michigan and Iowa, south to Georgia (Fig. 203).

3. A. hippocastanum L. Horse-Chestnut. A tree up to about 25 m. high, with large gummy varnished buds; fruit spiny. Introduced from Europe, sometimes seeding itself from cultivated trees (Fig. 204).

RHAMNUS L. (Rhamnaceae)

Deciduous shrubs or small trees. Twigs slender, sometimes spiny; pith continuous, white. Buds moderate, solitary, sessile.

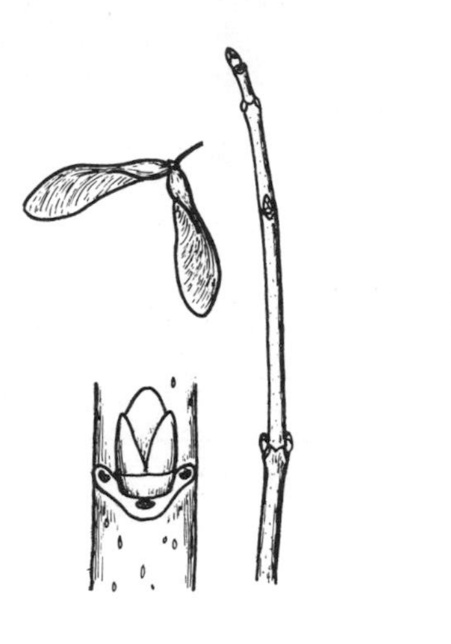

Fig. 200. Acer saccharinum

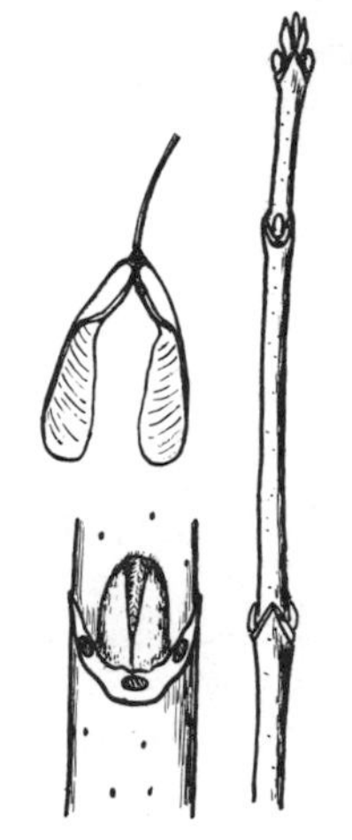

Fig. 201. Acer negundo

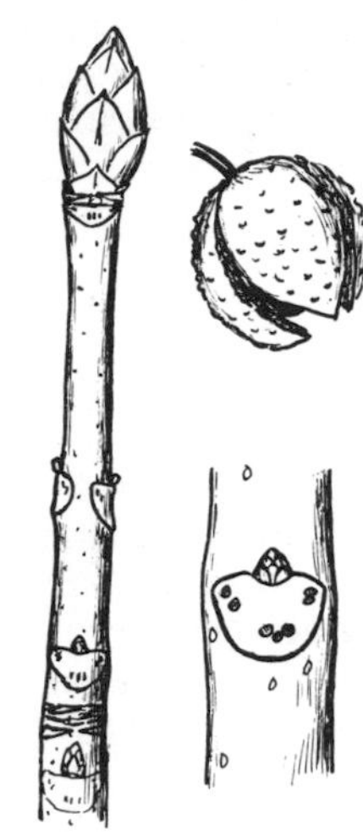

Fig. 202. Aesculus glabra

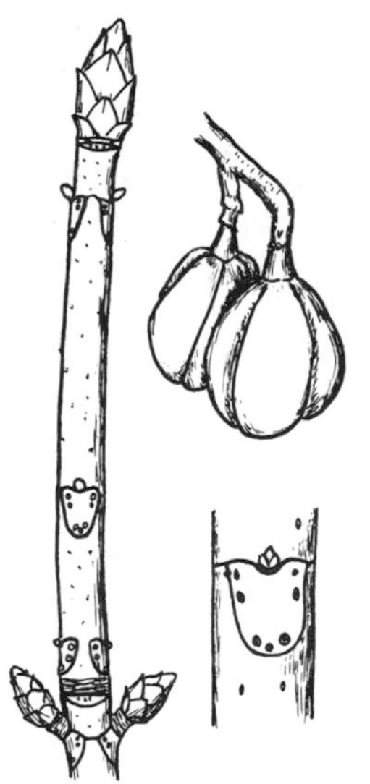

Fig. 203. Aesculus octandra

Fig. 204. Aesculus hippocastanum

Fig. 205. Rhamnus alnifolia

Leaf-scars alternate or sometimes nearly opposite, small, crescent-shaped or half-elliptical; bundle-traces 3 or joined in a series; stipule-scars minute.

Although the common name, Buckthorn is applied to species of this genus, many of them are not thorny.

a. Buds scaly

b. Twigs red or brown, glabrous — 1. R. alnifolia
b. Twigs gray, often downy — 2. R. lanceolata

a. Buds naked — 3. R. caroliniana

1. R. alnifolia L'Her. Alder-leaf Buckthorn. Low spreading shrub, 1.5-8 dm. high, with slender unarmed upright gray branches and smooth bark; twigs red or brown, glabrate; buds small, under 5 mm. long; leaf-scars alternate. Swamps and meadows, Newfoundland to British Columbia, south to West Virginia, Illinois, Nebraska, Wyoming and California (Fig. 205).

2. R. lanceolata Pursh. Lanceleaf Buckthorn. Erect shrub up to 2 m. high, with slender unarmed branches and smooth grayish bark; twigs reddish-brown, often downy; buds small, 5 mm. or less long. Thickets, Alabama to Texas, north to Pennsylvania and Nebraska (Fig. 206).

3. R. caroliniana Walt. Carolina Buckthorn. Indian Cherry. Shrub or small tree up to 11 m. high, without thorns; young branchlets puberulent; buds short, scarcely 5 mm. long; fruiting pedicels several in a cluster. Rich woods, Florida to Texas, north to Virginia, Missouri, and Nebraska (Fig. 207).

CEANOTHUS L. (Rhamnaceae)

Low deciduous shrubs. Twigs slender, rounded, puberulent; pith large, white, continuous. Buds small, solitary, sessile, ovoid. Leaf-scars alternate, half-elliptical, small; bundle-trace 1, transverse, sometimes compound; stipules small, persistent, or leaving narrow scars. The dry capsules and/or their bases are present in winter, like miniature cups-and-saucers.

1. C. americanus L. New Jersey Tea. Red-Root. Low unarmed shrub about 1 m. high, with slender upright branches; roots dark red; twigs slender, puberulent, green or brownish. Dry banks, Quebec to Manitoba, south to Florida and Texas (Fig. 208).

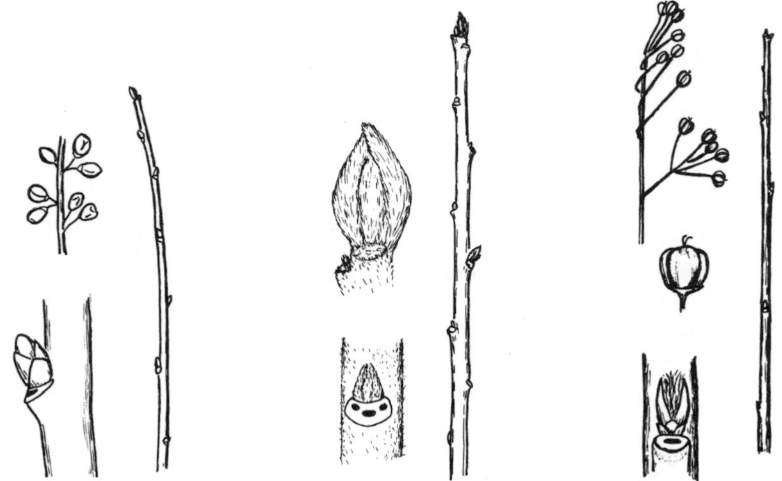

Fig. 206. Rhamnus lanceolata

Fig. 207. Rhamnus caroliniana

Fig. 208. Ceanothus americanus

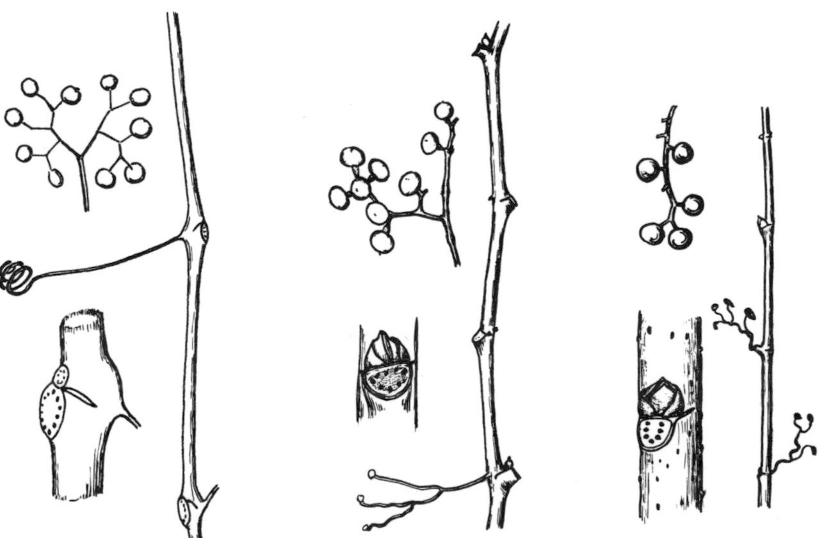

Fig. 209. Ampelopsis arborea

Fig. 210. Parthenocissus quinquefolia

Fig. 211. Parthenocissus tricuspidata

AMPELOPSIS Michx. (Vitaceae)

Soft-wooded or somewhat succulent deciduous climbers, bearing tendrils on the upper branches, opposite the leaf-scars. Stems somewhat angled; pith moderate, soon dividing into thin plates. Buds subglobose, solitary, sessile. Leaf-scars alternate, rounded; bundle-traces about a dozen, indistinct, in an ellipse; stipule-scars long and narrow.

1. A. arborea (L.) Koehne. Pepper-Vine. High-climbing or bushy nearly glabrous shrub; stems nearly terete; tendrils rather few. Swampy woods, Florida to Texas, north to Maryland, Missouri, and Oklahoma (Fig. 209).

PARTHENOCISSUS Planch. (Vitaceae)

Deciduous woody climbers, bearing tendrils opposite the leaf-scars, absent from every third node. Stems terete, moderate or slender; nodes swollen; pith relatively large, continuous. Buds moderate, frequently collaterally branched, sessile, round-conical; terminal bud absent. Leaf-scars indistinct, in an ellipse; stipule-scars long and narrow. Psedera Neck.

a. Branchlets usually pubescent, nearly terete — 1. P. quinquefolia
a. Branchlets essentially glabrous, channeled — 2. P. tricuspidata

1. P. quinquefolia (L.) Planch. Virginia Creeper. High-climbing or trailing woody plant; twigs usually pubescent; tendrils with 5-12 rather long branches ending in adhesive disks. Woods, Florida to Texas and Mexico, north to Maine and Minnesota (Fig. 210).

2. P. tricuspidata (Sieb. and Zucc.) Planch. Boston Ivy. High climbing; twigs glabrous; tendrils short, much-branched. Introduced from Asia, locally escaped from cultivation (Fig. 211).

VITIS L. (Vitaceae)

Hard-wooded deciduous climbers (or trailers), with usually very flaking bark. Stems striate, rounded or somewhat angled; tendrils opposite most of the leaf-scars. Buds subglobose, with 2 scales; terminal bud lacking. Leaf-scars alternate, half-round or crescent-shaped; bundle-traces several in a curved series, indistinct; stipule-scars long and narrow. Panicle-vestiges present in winter, and often the withered fruits also.

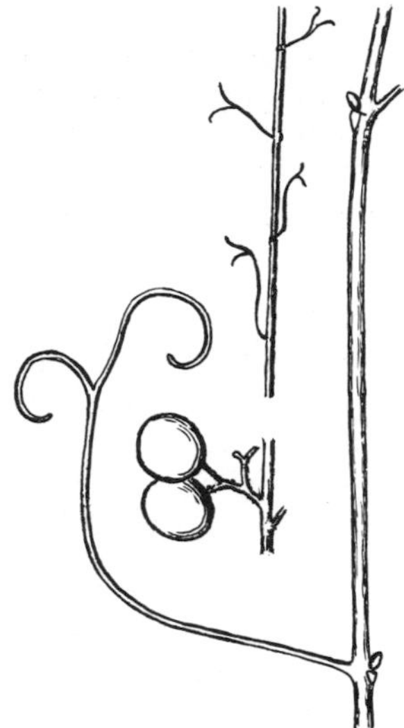
Fig. 212. Vitis labrusca

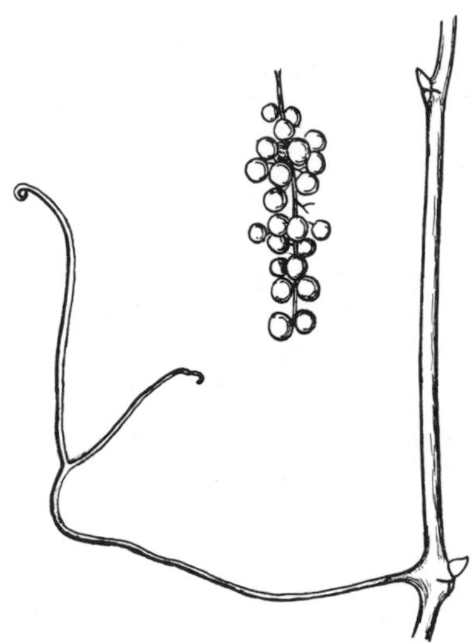
Fig. 213. Vitis aestivalis

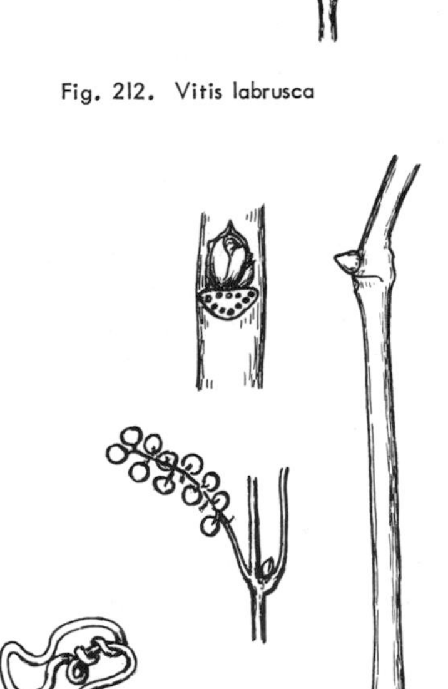
Fig. 214. Vitis argentifolia

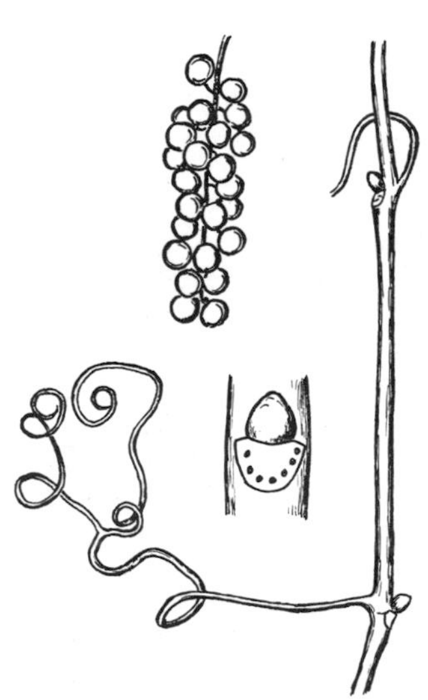
Fig. 215. Vitis riparia

a. Bark of main stem and branches shreddy and exfoliating, without distinct lenticels;pith with firm diaphragms at the nodes;tendrils, when present, forked

b. Branchlets bearing a tendril or inflorescence at each node — 1. V. labrusca

b. Tendrils (or inflorescences) intermittent, usually lacking at each third node

c. High-climbing lianas

d. Branchlets bluish-glaucous — 3. V. argentifolia

d. Branchlets not glaucous

e. Nodal diaphragms thin, 0.8-2 mm. thick — 4. V. riparia

e. Nodal diaphragms 2-6 mm. thick

f. Branchlets pubescent

g. Branchlets terete — 2. V. aestivalis

g. Branchlets angled — 6. V. baileyana

f. Branchlets glabrous — 5. V. vulpina

c. Low, bushy and spreading; tendrils absent, or only at tips of fruiting branches — 7. V. rupestris

a. Bark of main stem and branches close, not exfoliating; lenticels abundant;pith continuous, without diaphragms at the nodes; tendrils unbranched — 8. V. rotundifolia

1. V. labrusca L. Northern Fox Grape. High climbing or trailing, often ascending high trees, sometimes forming a stem 3.5 dm. in diameter; twigs and tendrils densely rusty-puberulent when young, becoming less so in winter; panicle little branched. Thickets and borders of woods, Maine to Michigan,. south to Kentucky, Tennessee, and Georgia (Fig. 212).

2. V. aestivalis Michx. Summer Grape. Pigeon Grape. A vigorous high-climbing vine with medium or short internodes and

Fig. 216. Vitis vulpina

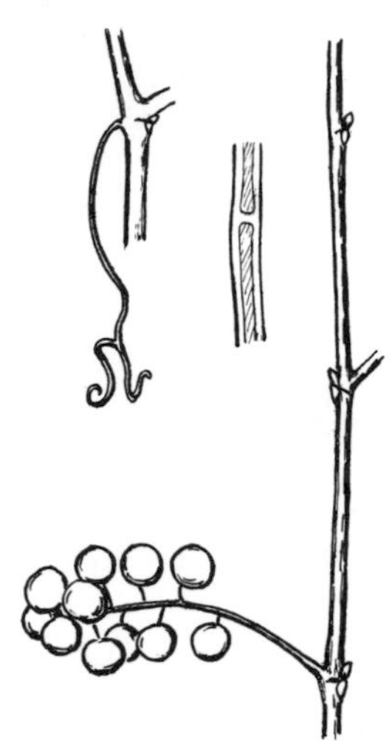

Fig. 217. Vitis rupestris

Fig. 218. Vitis rotundifolia

Fig. 219. Tilia americana

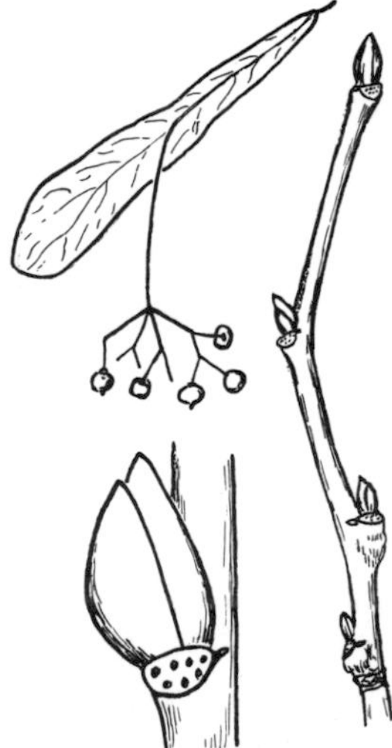

Fig. 220. Tilia heterophylla

pubescent twigs; panicle-vestiges compactly branched. Dry woods, Georgia to Texas, north to Massachusetts and Wisconsin (Fig. 213).

3. V. argentifolia Munson. Silverleaf Grape. (V. aestivalis var. argentifolia (Munson) Fernald; V. bicolor LeConte). High-climbing or long-trailing with simple tendrils; twigs with a bluish bloom, quite conspicuous when young, becoming less pronounced in winter. Dry woods, New Hampshire to Minnesota, south to Alabama and Kansas (Fig. 214).

4. V. riparia Michx. Riverbank Grape. (V. vulpina of authors, not L.). A strong climbing or trailing vine with glabrous branches; panicle vestiges compact. River-banks, Quebec to Manitoba and Montana, south to Tennessee, Missouri, Texas, and New Mexico (Fig. 215).

5. V. vulpina L. Frost Grape. Winter Grape. (V. cordifolia Michx.). Strong high climber, stem sometimes attaining a diameter of 3.5 dm. or more; tendrils forked; twigs glabrous; fruits black, becoming sweet after frost, persisting into winter. Bottomlands, Florida to Texas, north to New York, Illinois, Missouri, and Kansas (Fig 216).

6. V. baileyana Munson. Possum Grape. Rather slender, with short internodes and many short side branches; young shoots angled, covered for the first year with woolly hairs. Rich thickets, Virginia to Missouri, south to Alabama and Arkansas.

7. V. rupestris Scheele. Sand Grape. A shrub up to 2 m. high, spreading and rather bushy, sometimes slightly climbing; tendrils forked, few and small, or none. Sandy banks, Pennsylvania to Missouri, south to North Carolina, Arkansas, and Texas (Fig. 217).

8. V. rotundifolia Michx. Muscadine. Scuppernong. Trailing or high climbing, glabrous or nearly so throughout; tendrils simple, none opposite each third node. Woods, Florida to Texas, north to Delaware, Indiana, Missouri, and Oklahoma (Fig. 218).

TILIA L. (Tiliaceae)

Large deciduous trees. Twigs moderate, smooth, zigzag; pith continuous, pale. Buds solitary, obliquely sessile, rather large, inequilaterally ovoid, with about 2 green or red scales; terminal bud lacking. Leaf-scars alternate, 2-ranked, somewhat raised, half-elliptical; bundle-traces 3 or compound and scattered; stipule-scars unequal. Winter characters of the two principal native species are closely similar, but the withered leaves are usually present

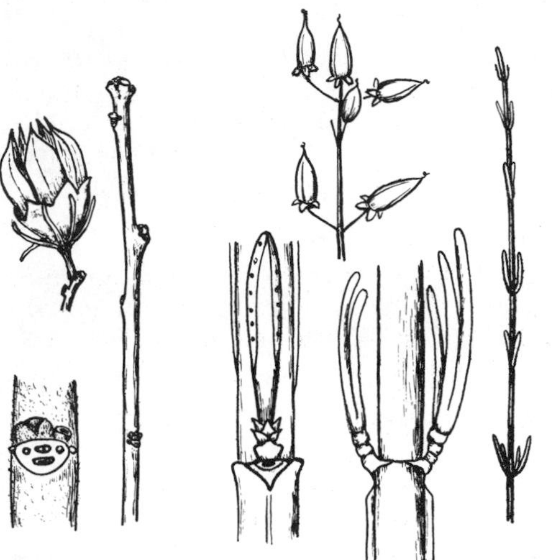
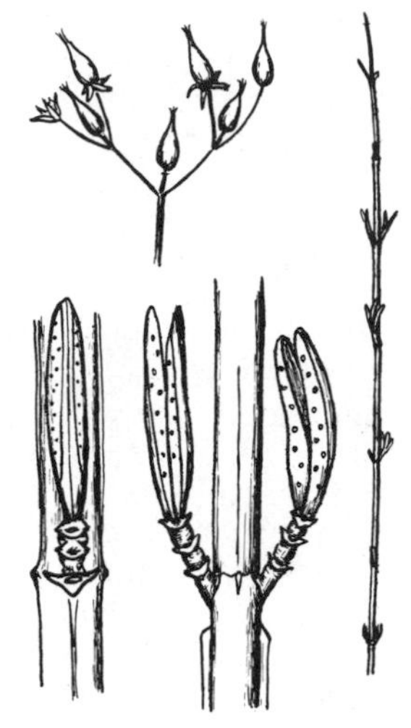

Fig. 221. Hibiscus syriacus

Fig. 222. Hypericum spathulatum

Fig. 223. Hypericum densiflorum

Fig. 224. Dirca palustris

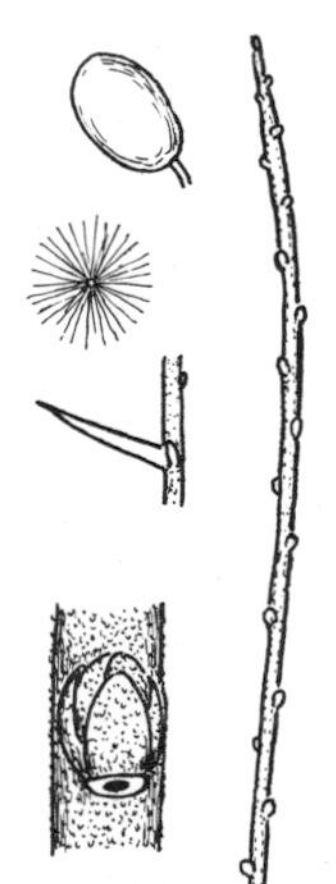

Fig. 225. Elaeagnus angustifolia

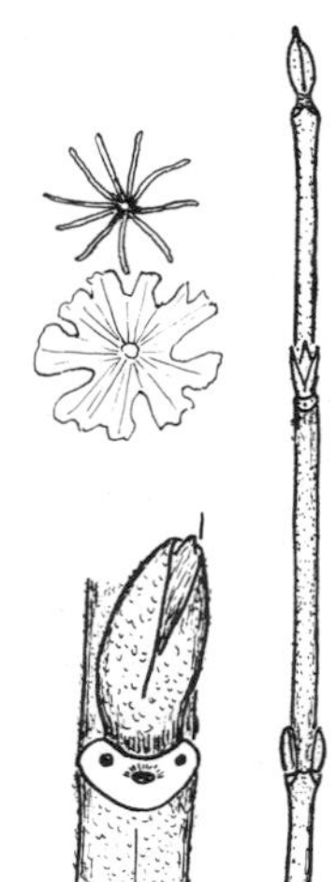

Fig. 226. Shepherdia argentea

and diagnostic.

a. Leaves essentially the same color on both surfaces; northern species	1. T. americana
a. Leaves whitened beneath; southern species	2. T. heterophylla

1. T. americana L. American Linden. Basswood. A large tree up to 40 m. high; bark gray, furrowed; twigs glabrous, green; fruit globose, 6-8 mm. in diameter. Rich woods, Quebec to Manitoba, south to Tennessee, Arkansas, and Texas (Fig. 219).

2. T. heterophylla Vent. White Basswood. White Linden. Linn. A large tree, 20-30 m. high, with a diameter of 1.5 m.; bark at first gray and smooth, becoming furrowed into flat ridges; twigs glabrous, reddish or yellowish-brown; fruit ellipsoid, 8 mm. long. Rich woods, Florida to Alabama and Missouri, north to West Virginia and New York (Fig. 220).

HIBISCUS L. (Malvaceae)

Small deciduous shrubs. Twigs slender, rounded, glabrescent; pith small, continuous, white, bordered with green. Buds not evident, covered by branch vestiges. Leaf-scars alternate, half-round, raised, decurrent in short ridges; bundle-traces about 4, indistinct; stipule-scars small.

1. H. syriacus L. Rose of Sharon. Shrubby Althaea. Shrub to 3 m. high, or more; twigs gray, at first villous, then glabrous; fruit a capsule 2.5 cm. long, persistent into winter. Introduced from Asia as a cultivated ornamental, established in roadsides (Fig. 221).

HYPERICUM L. (Hypericaceae)

Small deciduous shrubs (most species herbaceous) with flaking bark. Twigs slender, angled; pith relatively large, somewhat porous, finally excavated in older stems. Buds solitary, sessile, minute. Leaf-scars opposite, small, triangular; bundle-trace 1; stipule-scars lacking. Fruit a small capsule persistent in winter.

a. Dried capsule 7-15 mm. long	
b. Shrub 2-6 dm. high; capsule 7-10 mm. long	1. H. kalmianum
b. Shrub up to 2.5 m. tall; capsule 8-15 mm. long	2. H. spathulatum
a. Dried capsules 4-6 mm. long	3. H. densiflorum

1. H. kalmianum L. Kalm's St. John's-Wort. Slender shrub 2-6 dm. high, with papery whitish bark and ascending 4-edged branches, and 2-edged branchlets; corymb-vestiges open. Rocky soil, chiefly near the Great Lakes, Quebec to Ontario, south to New York and Illinois.

2. H. spathulatum (Spach) Steud. Shrubby St. John's Wort. (H. prolificum of authors, not L.). Shrub to 2.5 m. high, very bushy; twigs 2-edged. Dry or damp thickets, New York to Ontario and Minnesota, south to Georgia, Alabama, and Arkansas (Fig. 222).

3. H. densiflorum Pursh. Dense-flowered Shrubby St. John's-Wort. Shrub to 2 m. high, quite bushy; twigs slender, 2-edged. Wet acid soil, Florida to Texas, north to New Jersey, West Virginia, and Missouri (Fig. 223).

TAMARIX L. (Tamaricaceae)

Small trees and shrubs, deciduous by the fall of slender shoots bearing the small juniper-like leaves. Twigs slender, elongated; pith small, rounded, continuous. Buds small, sessile, compressed against the twig, solitary or multiple. Leaf-scars lacking, the twig-scars taking their place.

1. T. gallica L. Tamarisk. Glabrous shrub or small tree to 10 m. high; branches flexuous, red-brown; ultimate small branchlets falling with their leaves in autumn. Introduced from Europe and naturalized or casually escaped from cultivation.

HUDSONIA L. (Cistaceae)

Bushy heath-like low shrubs, covered with the small scale-like alternate evergreen pubescent leaves. Twigs very slender, terete; pith minute. Buds scarcely evident. Stipules lacking.

1. H. tomentosa Nutt. Beach-Heath. Hoary with villous tomentum; leaves 2 mm. long, appressed. Dunes and sandy places, Quebec to Alberta, south to Illinois, and on the coastal plain to North Carolina; also on high mountain summits, West Virginia.

DIRCA L. (Thymelaeaceae)

Low rounded deciduous shrubs with soft brittle wood but very tough bark (the name "leatherbark" would be more appropriate). Twigs rounded, glabrous, slender, light brown, or olive, enlarged upwards in each season's growth and swollen at the nodes; pith small, rounded, essentially continuous. Buds dark-silky, small, solitary, sessile, short-conical; terminal bud lacking. Leaf-scars alternate, nearly encircling the bud; bundle-traces 5 or more; stipule-scars none.

1. D. palustris L. Leatherwood. Shrub 1-2 m. high, with flexible branches and very tough bark; twigs forking, glossy. Rich woods, New Brunswick to Minnesota, south to Florida and Louisiana (Fig. 224).

ELAEAGNUS L. (Elaeagnaceae)

Shrubs or small trees, often spiny, stellate-hairy or with silvery or glistening-brown peltate scales; mostly deciduous. Twigs terete, rather slender; pith small, round, continuous. Buds small, solitary or collaterally branched in spine formation, exceptionally superposed, sessile, round, conical or oblong, with about 4 exposed scales. Leaf-scars alternate, half-round, minute; bundle-trace 1; stipule-scars lacking.

a. Branchlets covered with rusty scales — 1. E. commutata
a. Branchlets silvery — 2. E. angustifolia

1. E. commutata Bernh. Silverberry. (E. argentea Pursh). Stoloniferous unarmed shrub, 0.3-4 m. high, the younger branches covered with rusty scales. Dry limestone slopes, Quebec to Alaska, south to Minnesota, South Dakota and Utah.

2. E. angustifolia L. Oleaster. Russian Olive. Stoloniferous unarmed shrub to 4 m. high, the younger branches covered with silvery scales. Introduced from Eurasia, spreading somewhat from cultivation (Fig. 225).

SHEPHERDIA Nutt. (Elaeagnaceae)

Deciduous shrubs or small trees, often spiny, with glistening fringed silvery or red-brown peltate scales. Twigs subterete, slender; pith small, round, continuous. Buds rather small, solitary or multiple, stalked, oblong, with 2 or 4 valvate scales. Leaf-scars opposite, half-round, minute, slightly raised; bundle-trace 1; stipule-scars lacking.

a. Twigs red-brown-scurfy — 1. S. canadensis
a. Twigs silvery-scurfy — 2. S. argentea

1. S. canadensis L. Soapberry. Shrub 0.3-2 m. high; twigs with red-brown scurf. Rocky limestone slopes, Newfound-

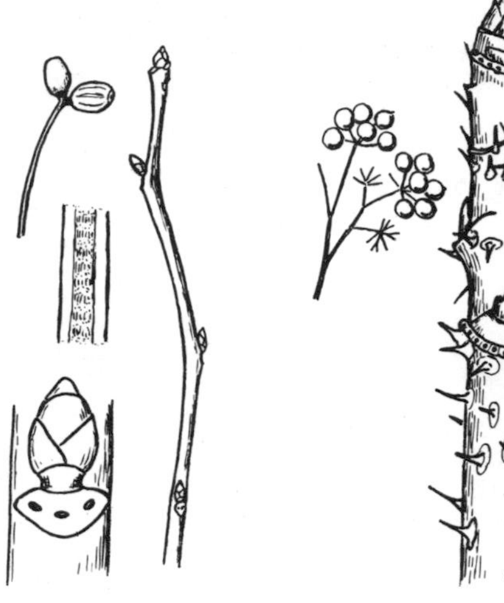

Fig. 227. Nyssa sylvatica

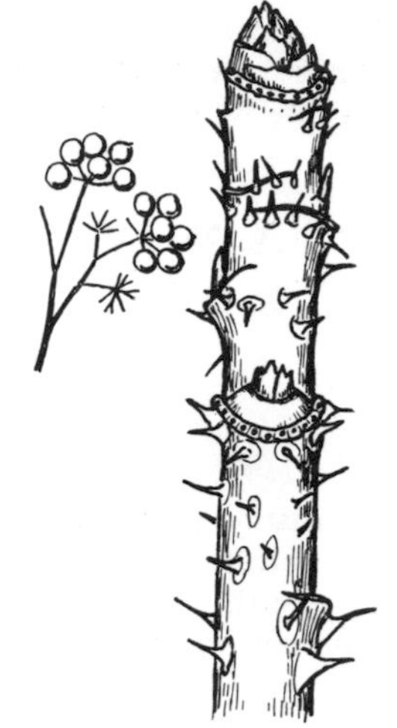

Fig. 228. Aralia spinosa

Fig. 229. Aralia hispida

Fig. 230. Hedera helix

Fig. 231. Cornus canadensis

Fig. 232. Cornus florida

land to Alaska, south to New York, Ohio, Minnesota, South Dakota and New Mexico.

2. S. argentea Nutt. Buffalo-berry. Shrub 1-6 m. high, somewhat thorny; twigs silvery-scurfy. Banks of streams, Manitoba to Alberta, south to Iowa, Kansas and New Mexico (Fig. 226).

NYSSA L. (Nyssaceae)

Tall deciduous trees. Twigs moderate, rounded; pith white, continuous, but with firmer plates at intervals. Buds sessile or slightly stalked, solitary or superposed, ovoid, with about 4 scales; terminal bud larger. Leaf-scars alternate, broadly crescent-shaped or triangular; bundle-traces 3; stipule-scars none. In some works Nyssa is placed in the Cornaceae.

1. N. sylvatica L. Black Gum. Sour Gum. A large tree reaching a height of 35-40 m. with a trunk diameter of 1.5 m.; bark on young stems smooth and grayish, on older trunks reddish-brown or grayish-black, scaly, and on very old trunks forming "alligator" bark; twigs glabrous; buds varicolored, soft-hairy. Rich woods, Maine to Missouri, south to Florida and Texas (Fig. 227).

ARALIA L. (Araliaceae)

Deciduous shrubs or small trees (many species herbaceous), sometimes with cortical prickles. Twigs often thick, rounded; pith large, continuous. Leaf-scars alternate, crescent-shaped, half-encircling the twigs; bundle-traces about 15, in a single series; stipule-scars none.

a. Tree or shrub, stem thick, prickly — 1. A. spinosa
a. Subshrub, stem moderate, bristly — 2. A. hispida

1. A. spinosa L. Hercules' Club. A shrub or tree, up to 10 m. high, with very prickly stems; branches gray or straw-colored, glabrous; umbels numerous, in compound panicles, persistent into winter. Rich woods, Florida to Texas, north to New Jersey, New York, Illinois and Iowa; escaped from cultivation farther north (Fig. 228).

2. A. hispida Vent. Bristly Sarsaparilla. Stem 2-9 dm. high, bristly, scarcely woody. Rocky woods and clearings, Newfoundland to Manitoba, south in the mountains to North Carolina (Fig. 229).

HEDERA L. (Araliaceae)

Woody evergreen climbers or trailers, with aerial roots as climbing organs. Stems moderate, terete; pith moderate, spongy. Buds small, conical, solitary, sessile, naked or with about 2 fleshy scales. Leaves alternate, palmately lobed and cordate, or lanceolate, ovate, or deltoid on older pendent shoots. Leaf-scars U-shaped, somewhat raised; bundle-traces 5 or 7; stipule scars lacking.

1. H. helix L. English Ivy. Leaves 3-5-lobed. Introduced from Europe, cultivated and becoming naturalized (Fig. 230).

CORNUS L. (Cornaceae)

Deciduous shrubs or small trees. Twigs slender, round;pith moderate, round, continuous or porous. Buds usually solitary, stalked or sessile, typically oblong. Leaf-scars usually opposite, crescent-shaped, commonly raised on petiole bases; bundle-traces 3; stipule-scars lacking, but the leaf-scars often meet or are joined by transverse lines which may resemble stipule-scars on cursory examination.

a. Low subshrub, 9-22 cm. high — 1. C. canadensis
a. Taller shrubs or trees

b. Leaf-scars opposite

c. Leaf-scars during the first winter on the ends of petiole-bases covering the leaf-buds; flower-buds biscuit-shaped — 2. C. florida
c. Leaf-buds not concealed by petiole-bases

d. Cyme-vestiges corymbose

e. Pith of 1-2-year old branchlets white

f. Branchlets red or brown, some of them prostrate and rooting at the tips (stoloniferous) — 3. C. stolonifera
f. Branchlets greenish, becoming pink in winter — 4. C. rugosa

e. Pith of 1-2-year old twigs brown or drab (rarely white in C. drummondii)

f. Branchlets gray — 5. C. drummondii

f. Branchlets red or purplish

g. Leaves broadly rounded at base — 6. C. amomum

g. Leaves tapered at base — 7. C. obliqua

d. Cyme-vestiges paniculate (not racemose, as the Latin name indicates) — 8. C. racemosa

b. Leaf-scars alternate (but crowded near the ends of the twigs, appearing as if whorled) — 9. C. alternifolia

1. C. canadensis L. Dwarf Cornel. Bunchberry. Almost herbaceous, but woody at the base; flowering stem from a nearly horizontal rootstock, erect, 9-22 cm. high; drupes red, globular, present in winter. Woods and openings, Greenland to Alaska, south to West Virginia, Illinois, Minnesota, South Dakota, New Mexico, and California (Fig. 231).

2. C. florida L. Flowering Dogwood. A shrub or small tree up to 12 m. high with grayish-brown "alligator" bark; twigs green or purple, glabrous; drupes ellipsoid, 1 cm. long, red (rarely yellow), persistent in winter. Acid woods, Florida to Texas and Mexico, north to Maine, Ontario, Missouri, and Kansas (Fig. 232).

3. C. stolonifera Michx. Red-Osier Dogwood. Erect shrub to 2.5 m. high, with some prostrate stoloniferous stems; branching loose and osier-like; twigs dark blood-red, glabrescent; pith white. Thickets, Newfoundland to Yukon, south to West Virginia, Iowa, Nebraska, New Mexico, Arizona, and California (Fig. 233).

4. C. rugosa Lam. Round-leaf Dogwood. (C. circinata L'Her.). Shrub 2-3 m. high, the branching rather loose; twigs green, becoming rather pink, bearing closely appressed pubescence; pith white; buds nearly sessile, hairy at the tip. Dry woods, Quebec to Manitoba, south to West Virginia, Illinois, and Iowa (Fig. 234).

5. C. drummondii Meyer. Roughleaf Dogwood. (C. asperifolia Michx.). Upright shrub or small tree up to 5 m. high; bran-

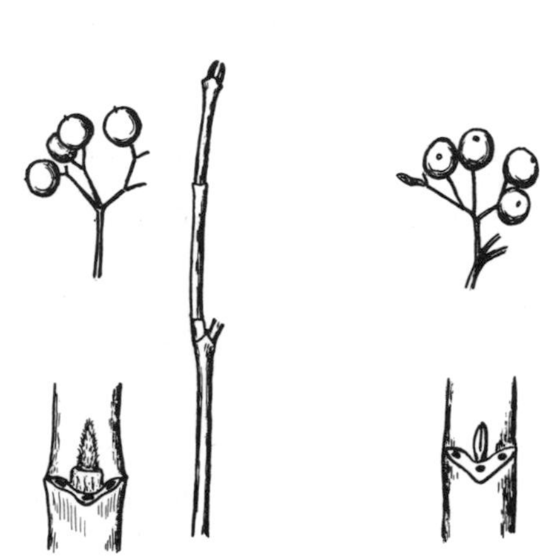

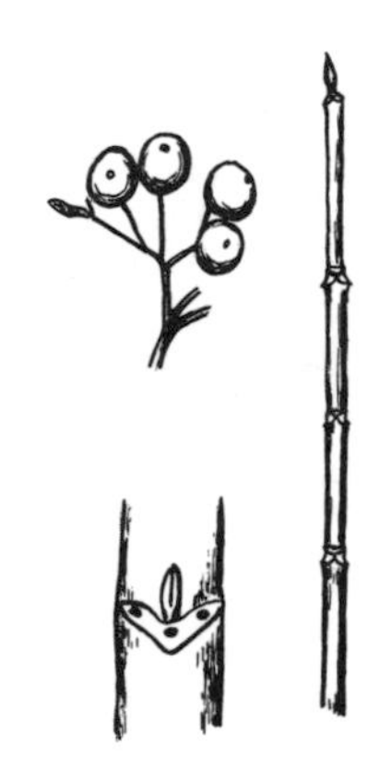

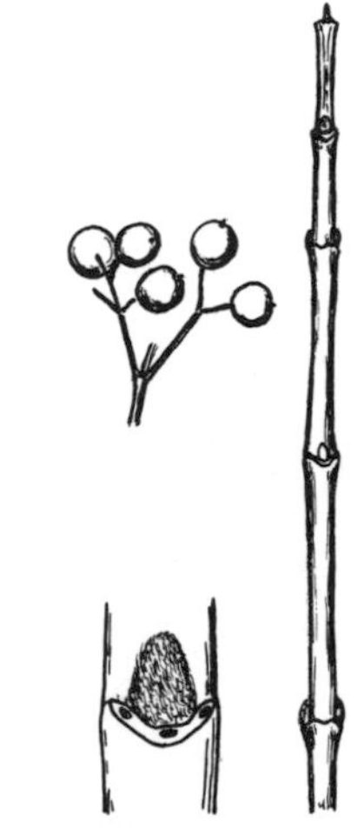

Fig. 233. Cornus stolonifera

Fig. 234. Cornus rugosa

Fig. 235. Cornus amomum

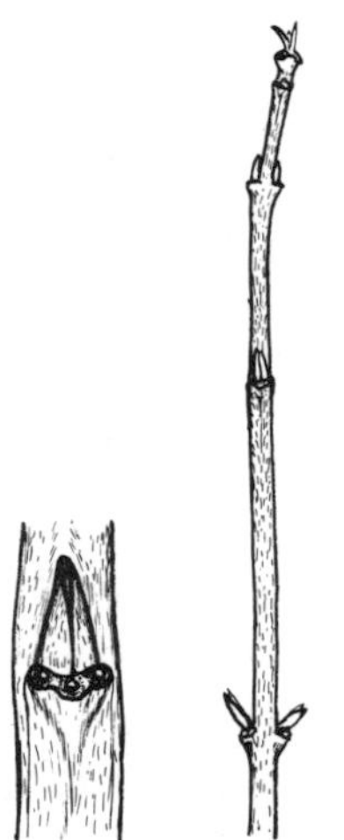

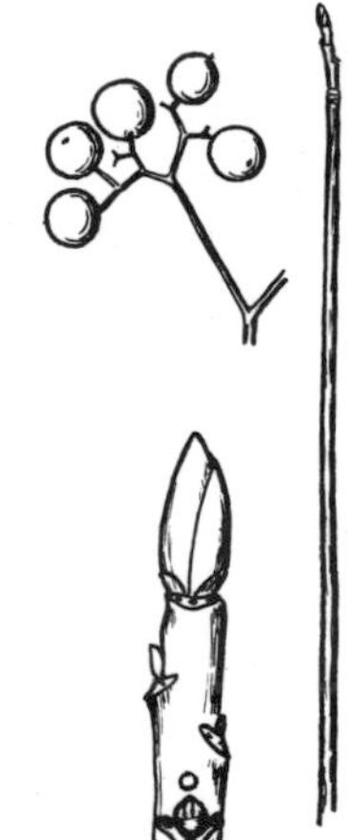

Fig. 236. Cornus obliqua

Fig. 237. Cornus racemosa

Fig. 238. Cornus alternifolia

ches gray, the young twigs reddish or purplish; pith slender, brown (rarely white). Shores, Mississippi to Texas, north to Ontario, Iowa, and Nebraska.

6. C. amomum Mill. Silky Cornel. Kinnikinnik. A shrub 1-3 m. high, with loose branching; not stoloniferous, although the lower branches may bend downwards and root in wet soil; twigs green, becoming purplish, bearing silky-downy or rusty pubescence; pith brown; buds nearly sessile. Damp thickets, Maine to Indiana, south to Georgia and Alabama (Fig. 235).

7. C. obliqua Raf. Silky Dogwood. Shrub or small tree to 3 m. high; branchlets purple or yellowish red. Damp thickets, New Brunswick to North Dakota, south to Kentucky, Arkansas and Oklahoma (Fig. 236).

8. C. racemosa Lam. Panicled Dogwood. (C. paniculata L'Her.). A shrub 2-5 m. high, with smooth light gray bark; twigs gray, glabrous; pith sometimes white but generally brownish, especially in 2-year-old branchlets. Dry soil, Maine to Ontario and Minnesota, south to Kentucky, Missouri, and Oklahoma (Fig. 237.).

9. C. alternifolia L. f. Alternate-leaf Dogwood. Shrub or small tree to 8 m. high, the branches spreading in irregular whorls to form horizontal tiers; branchlets glabrous, greenish; leaf-scars alternate, but crowded. Dry woods, Newfoundland to Ontario and Minnesota, south to Georgia, Alabama, and Missouri (Fig. 238).

CLETHRA L. (Clethraceae)

Deciduous shrubs or small trees with stellate pubescence. Twigs rounded; pith relatively large, continuous, with a network of firmer strands. Buds solitary, sessile, the lateral small, the terminal larger, rosy, acute. Leaf-scars alternate, clustered toward the tip, triangular; bundle-trace 1; stipule-scars none.

a. Shrub to 3 m. high, on the coastal plain	1. C. alnifolia
a. Taller shrub or small tree, to 6 m. high, in the mountains	2. C. acuminata

1. C. alnifolia L. Sweet Pepperbush. Shrub 1-3 m. high; twigs slender; terminal bud about 5 mm. long. Damp woods, on the coastal plain, Texas to Florida, north to Maine (Fig. 239).

Fig. 239. Clethra alnifolia Fig. 240. Clethra acuminata Fig. 241. Ledum groenlandicum

Fig. 242. Rhododendron maximum

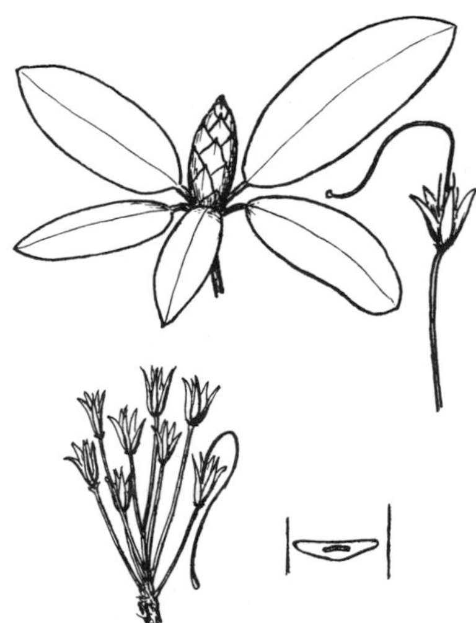

Fig. 243. Rhododendron catawbiense

2. C. acuminata Michx. White-Alder. A shrub or small tree to 6 m. high, with upright or spreading branches; twigs slender; buds tomentulose, the terminal 5 mm. long; fruit a persistent ovoid nodding capsule, 5 mm. long. Woods, in the mountains, West Virginia to Georgia and Tennessee (Fig. 240).

LEDUM L. (Ericaceae)

Evergreen bog shrubs. Twigs rather slender, rounded; pith small, somewhat 3-sided, spongy, brownish. Buds solitary, sessile, somewhat compressed, small, with about 3 exposed scales; the terminal inflorescence buds large, round or ovoid, with about 10 broad glandular-dotted scales. Leaves alternate, simple, entire, elliptical or oblong. Leaf-scars half-elliptical or cordate, the lowest linear; bundle-trace 1; stipule-scars lacking.

1. L. groenlandicum Oeder. Labrador-Tea. Erect shrub, 1 m. or less high; leaves evergreen, oblong or linear-oblong, entire, revolute, clothed with rusty wool underneath, 2-5 cm. long. Bogs and acid soil, Greenland to Alaska, south to Pennsylvania, Wisconsin, Minnesota, Alberta and Washington (Fig. 241).

RHODODENDRON L.* (Ericaceae)

Evergreen or deciduous shrubs or small trees. Twigs terete; pith small, roundish, continuous. Buds solitary, sessile, the upper ovoid with 6 or more ciliate scales and the flower-buds usually much enlarged, the lower successively smaller. Leaf-scars alternate, shield-shaped or the lowest linear; bundle-trace 1 or 3 or many; stipule-scars none. Fruit a dry capsule, often present in winter. The deciduous species have been placed by some authors in the genus Azalea L.

a. Leaves evergreen

b. Leaves elongate-oblanceolate, the base acute — 1. R. maximum
b. Leaves broadly elliptical, the base rounded — 2. R. catawbiense

a. Leaves deciduous

b. Buds essentially glabrous

c. Twigs entirely glabrous — 8. R. arborescens
c. Twigs usually bearing sparse long hairs

*Mr. E. L. Manigault has assisted materially in the preparation of this key.

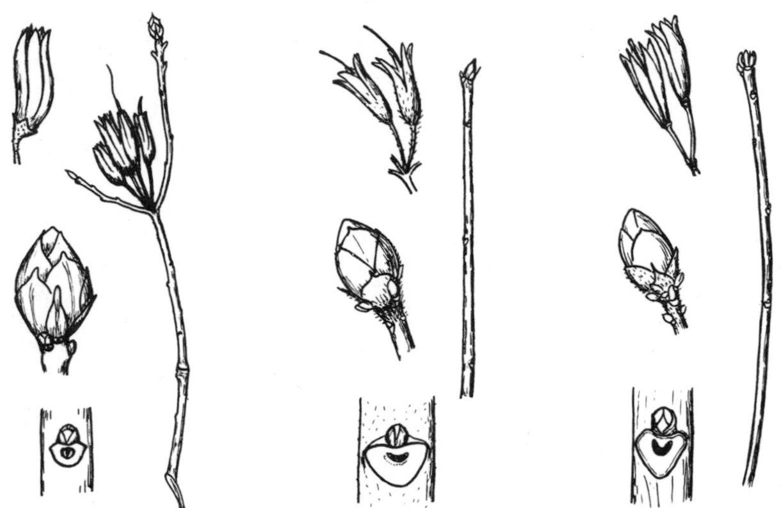

Fig. 244. Rhododendron canadense

Fig. 245. Rhododendron calendulaceum

Fig. 246. Rhododendron nudiflorum

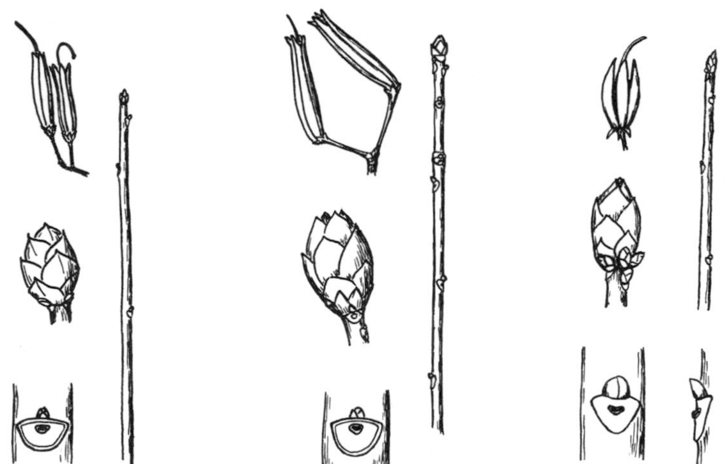

Fig. 247. Rhododendron roseum

Fig. 248. Rhododendron viscosum

Fig. 249. Rhododendron arborescens

d. Buds brown	4. R. calendulaceum
d. Buds rosy	5. R. nudiflorum
b. Buds pubescent	
c. Low shrub to 1 m. high	3. R. canadense
c. Much taller shrubs	
d. Twigs reddish	4. R. calendulaceum
d. Twigs buff or gray	
e. Twigs tomentulose, at least near the tip	6. R. roseum
e. Twigs sparsely strigose-hirsute	
f. Pedicels glandless or nearly so	5. R. nudiflorum
f. Pedicels copiously glandular	7. R. viscosum

1. R. maximum L. Rhododendron. Great Laurel. Large evergreen shrub or small tree to 12 m. high, the young branchlets pubescent; leaves 10-20 cm. long, thick, acute at both ends. Moist woods, mostly in the mountains, Georgia to West Virginia and Nova Scotia (Fig. 242).

2. R. catawbiense Michx. Mountain Rose Bay. Spreading evergreen shrub or small tree 2-6 m. high; leaves oval or oblong, rounded at both ends, 5-15 cm. long. Rocky slopes, in the mountains, Georgia and Alabama north to West Virginia (Fig. 243).

3. R. canadense (L.) Torr. Rhodora. Shrub to 1 m. high, with strongly ascending branches. Bogs and barrens, Newfoundland to Quebec, south to Pennsylvania (Fig. 244).

4. R. calendulaceum (Michx.) Torr. Flame Azalea. Yellow Honeysuckle. Deciduous shrub to 3 m. high, the twigs usually tomentulose, at least near the tips, leaf buds brown, glabrous except for ciliate scales, the flower buds greenish; capsule linear-oblong, more or less pubescent, 1.5-2 cm. long. Open woods, mostly in the mountains, Georgia and Alabama, north to Pennsylvania and Ohio (Fig. 245).

Fig. 250. Menziesia pilosa

Fig. 251. Kalmia latifolia

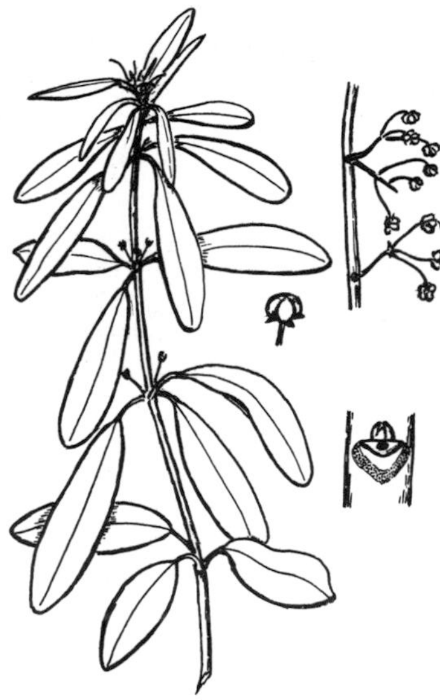

Fig. 252. Kalmia angustifolia

Fig. 253. Kalmia polifolia

5. R. nudiflorum (L.) Torr. Pink Azalea. Pink Honeysuckle. Deciduous shrub to 2 m. high or sometimes taller; twigs buff or gray, sparsely long-hairy or glabrate; buds glabrous, glandular or puberulous; capsules 1-2 cm. long with ascending long hairs, not glandular. Woods, South Carolina and Tennessee to Massachusetts, New York, and Ohio (Fig. 246).

6. R. roseum (Loisel.) Rehd. Mountain Azalea. Hoary Azalea. A deciduous branching shrub 1-5 m. high with buff or gray finely pubescent twigs; buds puberulous; capsules glandular, 1.2-1.8 cm. long. Dry thickets, mostly in the mountains, Maine to Quebec, south to Virginia and Missouri (Fig. 247).

7. R. viscosum (L.) Torr. Clammy Azalea. Swamp Honeysuckle. Deciduous shrub to 3 m. high or sometimes taller, the young twigs buff or gray, bristly with long hairs, becoming glabrate; buds puberulous; capsule 1-2 cm. long, glandular. Swamps, mostly in the mountains, Maine to Ohio, south to South Carolina and Tennessee (Fig. 248).

8. R. arborescens (Pursh) Torr. Smooth Azalea. White Honeysuckle. Spreading deciduous shrub or small tree, 2-6 m. high; twigs usually entirely glabrous; buds and pedicels glabrous except for occasional stalked glands; capsules oblong, densely glandular, 1.2-1.6 cm. long. Swamps and stream-banks, mostly in the mountains, Pennsylvania to Georgia and Alabama (Fig. 249).

MENZIESIA Sm. (Ericaceae)

Deciduous shrubs with shredding bark. Twigs slender; pith small, continuous. Buds solitary, sessile, ovoid, small, the terminal and flower buds larger. Leaf-scars crowded toward the tips of the branches, small, triangular or linear; bundle-trace 1; stipule-scars none. Fruit an ovoid capsule persistent in winter.

1. M. pilosa (Michx.) Juss. Alleghany Menziesia. Shrub to 2 m. high, with slightly pubescent twigs; capsule about 4 mm. high, bristly-glandular. Mountain woods, in the Alleghenies from Pennsylvania to Georgia (Fig. 250).

KALMIA L. (Ericaceae)

Evergreen shrubs or small trees. Twigs moderate or slender; pith small, continuous. Buds minute, solitary, sessile; terminal bud abortive. Leaves alternate (or seemingly opposite or in whorls of 3), entire. Leaf-scars half-round or shield-shaped; bundle-trace a transverse line; stipule-scars none. Fruit a small globose cap-

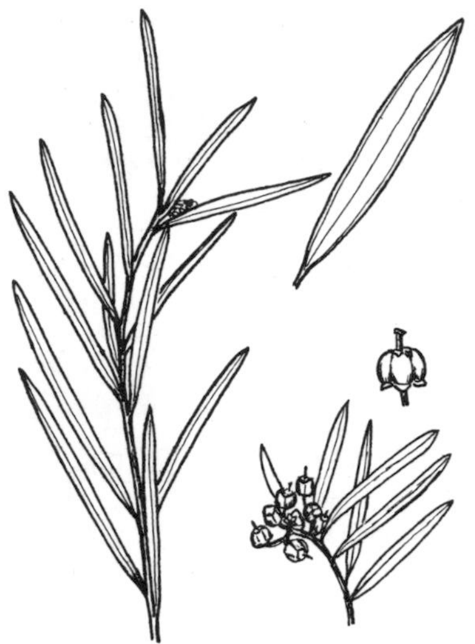

Fig. 254. Andromeda glaucophylla

Fig. 255. Pieris floribunda

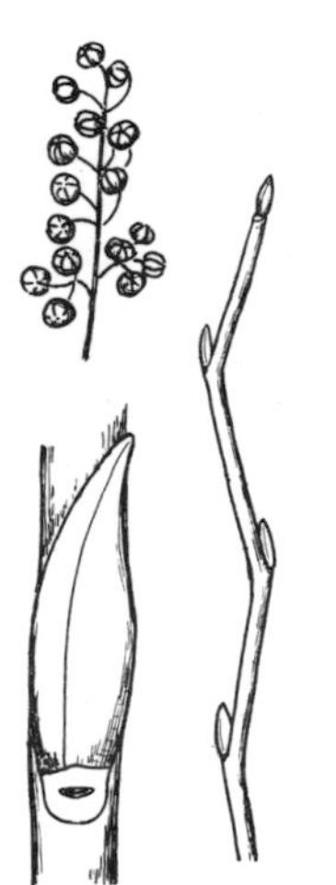

Fig. 256. Lyonia ligustrina

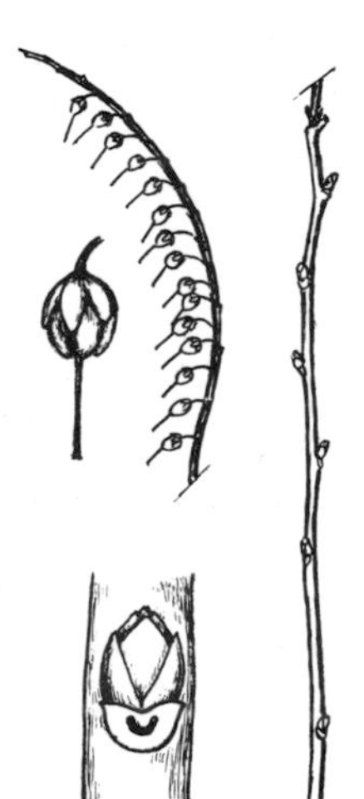

Fig. 258. Oxydendrum arboreum

Fig. 257. Leucothoë recurva

sule persistent in winter.

a. Large shrubs or small trees — 1. K. latifolia
a. Low shrubs

b. Twigs terete — 2. K. angustifolia
b. Twigs 2-edged — 3. K. polifolia

1. K. latifolia L. Mountain Laurel. Calico-Bush. Shrub or small tree, up to 10 m. high, generally much smaller; twigs glabrous; leaves elliptic to elliptic-lanceolate, 5-10 cm. long, acute or short-acuminate; capsule glandular, 5-7 mm. in diameter. Rocky woods, Florida to Louisiana, north to New England, Ohio, and Indiana (Fig. 251).

2. K. angustifolia L. Lambkill. Sheep-Laurel. Slender shrub up to 1.7 m. high; branches terete, strongly ascending; leaves evergreen, opposite or in whorls of three, flat, thin, oblong to elliptic-lanceolate, glabrous or puberulent, ferruginous when young. Acid soil, Labrador to Manitoba, south to Virginia, Georgia and Michigan (Fig. 252).

3. K. polifolia Wang. Pale-Laurel. Bog-Laurel. Slender straggling shrub to 7 dm. high, with 2-edged branches; leaves evergreen, opposite or in threes, firm, lanceolate or linear, 0.7-3.5 cm. long, lustrous-green above, conspicuously whitened beneath. Peat bogs, Labrador to Alaska, south to Pennsylvania, Minnesota, Idaho, and Oregon (Fig. 253).

ANDROMEDA L. (Ericaceae)

Evergreen bog shrubs. Twigs slender, somewhat 3-sided; pith small, 3-sided, continuous. Buds small, solitary, sessile, ovoid, subglobose or conical. Leaves alternate, simple, entire. Leaf-scars small, half-round or triangular; bundle-trace 1; stipule-scars none. Fruit a globose capsule persistent in winter.

1. A. glaucophylla Link. Bog Rosemary. Low shrub to 1 m. high, usually little branched; leaves linear or linear-oblong, prominently whitened beneath, 2.5-6 cm. long, the margins strongly revolute; capsule 4 mm. long. Bogs, Greenland to Manitoba, south to West Virginia, Indiana, and Minnesota (Fig. 254).

PIERIS D. Don (Ericaceae)

Evergreen shrubs or small trees. Twigs slender; pith obscurely 3-sided, continuous. Buds small, globose, the terminal one lacking. Leaves alternate, lanceolate-oblong. Leaf-scars small, half-round; bundle-trace 1; stipule-scars none. Fruit a globose capsule, present in the winter along with the terminal and axillary racemes or panicles of next season's flowers.

1. P. floribunda (Pursh) B. and H. Mountain Fetterbush. (Andromeda floribunda Pursh). Erect shrubs to 2 m. high; leaves elliptic-ovate to oblong-lanceolate, 3-8 cm. long, acute or acuminate, crenate-serrulate, ciliate; capsule globose-ovoid, 5-6 mm. long. Moist soil, in the Alleghenies, West Virginia to Georgia (Fig. 255).

LYONIA Nutt. (Ericaceae)

Deciduous or evergreen shrubs. Twigs slender, somewhat 3-sided; pith round. Buds solitary, sessile, oblong, appressed, with 2 visible scales. Leaf-scars alternate, small, half-round; bundle-trace 1; stipule-scars none. Fruit a globular capsule persistent in winter.

1. L. ligustrina (L.) DC. Maleberry. Deciduous much-branched shrub to 4 m. tall; twigs puberulent or glabrous; capsule about 3 mm. in diameter. Usually in moist thickets, Florida to Texas, north to New England, Kentucky and Oklahoma (Fig. 256).

LEUCOTHOE D. Don (Ericaceae)

Evergreen or deciduous shrubs. Twigs rather slender, subterete; pith roundish or somewhat 3-sided, continuous. Buds small, solitary, sessile, globose or ovoid, with 3 or 4 exposed scales, the end-bud lacking. Leaf-scars alternate, small, crescent-shaped or half-round; bundle-trace 1; stipule-scars lacking. Capsules small, depressed-globose, often present in winter.

a. Leaves evergreen — 1. L. editorum
a. Leaves deciduous
  b. Capsules deeply 5-lobed; racemes curved — 2. L. recurva
  b. Capsules not lobed; racemes straight — 3. L. racemosa

1. L. editorum (Fernald and Schub.) Dog-Hobble. Switch-Ivy. (L. catesbaei Gray). Shrub to 2 m. high, with loosely spreading branches; leaves evergreen, ovate-lanceolate, pointed, closely serrulate, up to 1.5 dm. long. Moist woods, along the mountains from Virginia to Georgia.

2. L. recurva (Buckley) Gray. Fetterbush. A widely-branched deciduous shrub, 0.8-3 m. high; capsules 4 mm. in diameter, deeply 5-lobed; racemes curved. Dry soil, mountains from Virginia and West Virginia to Alabama (Fig. 257).

3. L. racemosa (L.) Gray. Fetterbush. Shrub up to 4 m. high, with ascending branches; capsules scarcely lobed; racemes straight. Moist thickets, mostly on the coastal plain, Florida to Louisiana, north to Massachusetts.

OXYDENDRUM DC. (Ericaceae)

Small or medium-sized deciduous tree. Twigs slender, zigzag; pith pale, continuous. Buds rather small, conical-globose, solitary, sessile, with about 6 scales; terminal bud lacking. Leaf-scars alternate, small, half-round; bundle-trace 1, curved; stipule-scars none. Small capsules (in one-sided racemes clustered in panicles) present in winter.

1. O. arboreum (L.) DC. Sourwood. A tree to 25 m. high but usually much smaller; bark deeply fissured; twigs glabrous or sparingly puberulous, olive or bright red; capsule 5 mm. long. Acid woods, Florida to Louisiana, north to Pennsylvania and Indiana (Fig. 258).

CHAMAEDAPHNE Moench. (Ericaceae)

Evergreen bog shrub. Twigs slender, roundish, at first puberulent and scurfy, then with shredding gray bark, finally smooth and deep red-brown; pith small, roundish, continuous. Buds solitary, sessile, small, globose, with about 3 exposed scales. Leaves simple, alternate, entire, scurfy beneath. Leaf-scars minute, crescent-shaped; bundle-trace 1; stipule-scars lacking. Capsule small, depressed-globose, present in winter.

1. C. calyculata (L.) Moench. Leatherleaf. Shrub up to 1 m. high or taller; leaves evergreen, coriaceous, scurfy, especially beneath, 2.5-5 cm. long. Bogs, Labrador to Alaska, south to Ohio, Wisconsin, Iowa, Alberta, and British Columbia; also in Eurasia (Fig. 259).

EPIGAEA L. (Ericaceae)

Evergreen prostrate and trailing scarcely shrubby plants with exfoliating bark. Twigs slender; pith moderate, rounded, continuous. Buds solitary or multiple, hairy. Leaves alternate, elliptic-ovate, cordate, entire, ciliate, reticulate.

Fig. 259. Chamaedaphne calyculata

Fig. 260. Epigaea repens

Fig. 261. Gaultheria procumbens

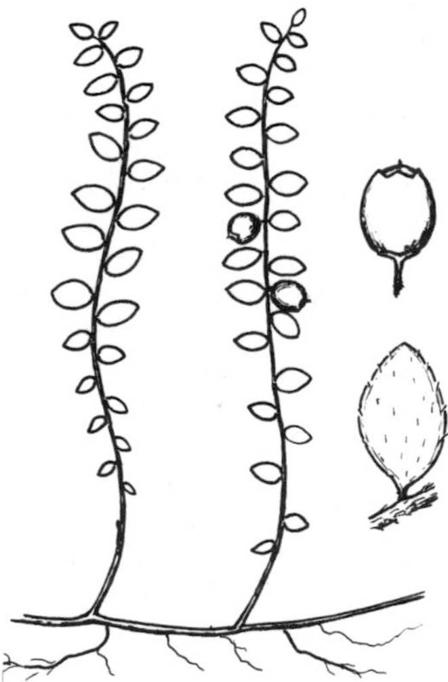

Fig. 262. Gaultheria hispidula

1. E. repens L. Trailing Arbutus. Branches red-bristly, 1.5-4 dm. high; leaves 2.5-7.5 cm. long; capsules depressed-globular, 5-lobed, 8-10 mm. in diameter. Sandy woods, Florida to Mississippi, north to Labrador and Saskatchewan (Fig. 260).

GAULTHERIA L. (Ericaceae)

Evergreen aromatic erect or trailing small shrubs. Twigs moderate or very slender, rounded; pith small. Buds solitary, sessile, ovoid, minute. Leaves alternate, simple, obovate or ovate, entire. Fruit a fleshy globular capsule resembling a berry, present in winter.

a. Fruiting stem erect, 5-15 cm. high, leaves 1.5-5 cm. long — 1. G. procumbens
a. Stems trailing and creeping; leaves 0.5-1 cm. long — 2. G. hispidula

1. G. procumbens L. Mountain-tea. Teaberry. Spicy Wintergreen. Stems slender, trailing on the ground or subterranean; leaves 2.5-5 cm. long; branches ascending, 5-15 cm. high; berry depressed, red, 8-12 mm. in diameter, mealy, spicy. Sterile woods, Newfoundland to Manitoba, south to Georgia, Alabama, Wisconsin, and Minnesota (Fig. 261).

2. G. hispidula (L.) Bigel. Creeping Snowberry. (Chiogenes hispidula T. and G.). Stems 1-3 dm. long, creeping and branching; leaves 4-10 mm. long; berry white, globular, aromatic, 6 mm. across. Mossy woods and bogs, Labrador to British Columbia, south to Pennsylvania, North Carolina, Michigan, Minnesota, and Idaho (Fig. 262).

ARCTOSTAPHYLOS Adans. (Ericaceae)

Evergreen shrubs. Twigs slender, somewhat 3-sided or 5-sided; pith small, slightly angled, continuous. Buds solitary, sessile, ovoid, with about 3 exposed scales. Leaves alternate, spatulate, rather small. Leaf-scars small, crescent-shaped; bundle-trace 1; stipule-scars lacking.

1. A. uva-ursi (L.) Spreng. Bearberry. Trailing shrub with long flexible branches covered with papery reddish to ashy exfoliating bark; twigs minutely tomentulose-viscid, becoming glabrate; leaves evergreen, obovate to spatulate, coriaceous, entire, 0.9-3 cm. long. Exposed rocky or sandy soil, Greenland to Alaska, south to Virginia, Illinois, Colorado, and California (Fig. 263).

Fig. 263. Arctostaphylos uva-ursi

Fig. 264. Gaylussacia brachycera

Fig. 265. Gaylussacia dumosa

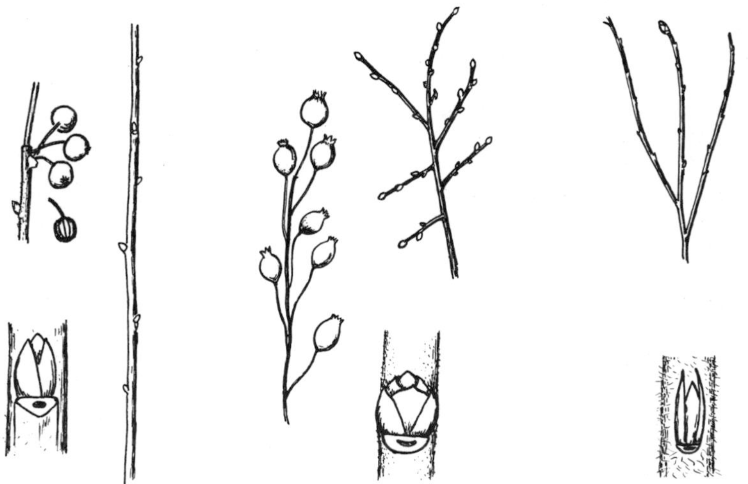

Fig. 266. Gaylussacia baccata

Fig. 267. Vaccinium stamineum

Fig. 268. Vaccinium myrtilloides

GAYLUSSACIA HBK (Ericaceae)

Low deciduous or evergreen shrubs. Twigs slender, 3-sided or roundish; pith small, continuous. Buds solitary, sessile, ovoid, small; terminal bud lacking. Leaf-scars alternate, 3-sided; bundle-trace 1; stipule-scars none.

a. Leaves evergreen — 1. G. brachycera
a. Leaves deciduous

b. Buds with 4 or 5 exposed scales; mostly on the coastal plain — 2. G. dumosa
b. Buds with 2 or 3 exposed scales; northern, mostly in the mountains southwards — 3. G. baccata

1. G. brachycera (Michx.) Gray. Box Huckleberry. A shrub 1 m. high, with creeping and ascending stem and spreading glabrous angled branches; leaves evergreen, serrulate, 1.5-2.5 cm. long. Sandy woods, local, Delaware to West Virginia and Tennessee (Fig. 264).

2. G. dumosa (Andr.) T. and G. Dwarf Huckleberry. A low shrub with creeping stem and erect branches to 0.5 m. high; twigs puberulent. Dry barrens, mostly on the coastal plain, Florida to Mississippi, north to Newfoundland (Fig. 265).

3. G. baccata (Wang.) K. Koch. Black Huckleberry. Upright shrub about 1 m. high, with the young growth densely resinous; twigs more or less pubescent. Thickets and woods, Newfoundland to Saskatchewan, south to Georgia and Louisiana (Fig. 266).

VACCINIUM L. (Ericaceae)

Deciduous shrubs, usually under 1 m. tall, sometimes trailing or arborescent (some species evergreen). Twigs slender, more or less angled; pith small, continuous. Buds small, solitary, sessile, with 2 or more scales; terminal bud deciduous. Leaf-scars alternate, small, half-round; bundle-trace 1; stipule-scars none.

a. Leaves deciduous

b. Buds oblong, appressed, with 2 obtuse exposed scales — 6. V. erythrocarpum
b. Buds ovoid or globose; scales several or pointed

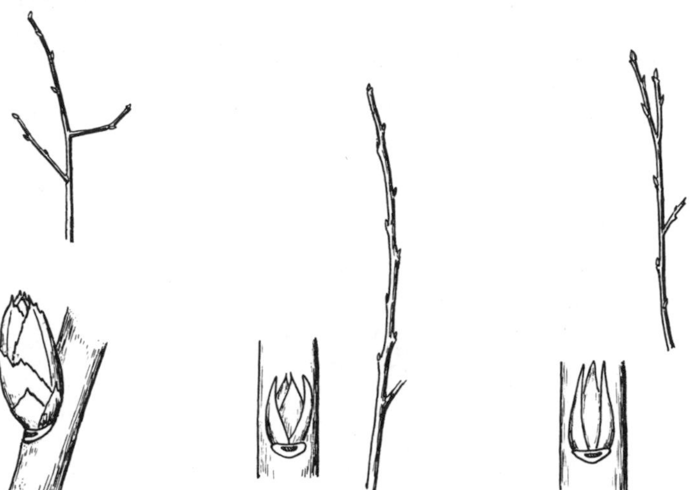

Fig. 269. Vaccinium vacillans

Fig. 270. Vaccinium angustifolium

Fig. 271. Vaccinium corymbosum

Fig. 272. Vaccinium erythrocarpum

Fig. 273. Vaccinium oxycoccos

Fig. 274. Vaccinium macrocarpon

c. Buds ovoid or oblong, ascending or appressed

d. Low shrubs, 0.5-9 dm. high

e. Twigs nearly terete, very hairy — 2. V. myrtilloides

e. Twigs grooved above the buds or angular

f. Twigs smooth — 4. V. angustifolium

f. Twigs granular-warty — 3. V. vacillans

d. Taller shrubs, up to 4 m. high — 5. V. corymbosum

c. Buds subglobose or ovoid, divergent — 1. V. stamineum

a. Leaves evergreen; stems trailing

b. Leaves very small, 4-10 mm. long,revolute — 7. V. oxycoccos

b. Leaves larger, 6-18 mm. long, not revolute — 8. V. macrocarpon

1. V. stamineum L. Deerberry. Buckberry. Squaw Huckleberry. A diffusely branched shrub 3-9 dm. high, with pubescent twigs; buds subglobose. Dry thickets, Florida to Louisiana, north to Massachusetts, Ontario, and Missouri (Fig. 267).

2. V. myrtilloides Michx. Canada Blueberry. (V. canadense Kalm). A low shrub 2-6 dm. high; twigs nearly terete, very hairy; buds ovoid or oblong, scales several. Moist soil, Newfoundland and Quebec to British Columbia, south to Virginia, Ohio, Iowa, Saskatchewan and Montana (Fig. 268).

3. V. vacillans Torr. Late Low Blueberry. Low shrub, 3-9 dm. high; twigs glabrous, yellowish-green, distinctly warty, angled; buds ovoid or oblong. Dry soil, Georgia to Missouri,north to Nova Scotia, Ohio, and Iowa(Fig. 269).

4. V. angustifolium Ait. Low Sweet Blueberry. Early Low Blueberry. A low shrub to 2 dm. high; twigs smooth. Open barrens, Labrador and Quebec to Minnesota, south to Iowa, Illinois, and West Virginia (Fig. 270).

5. V. corymbosum L. Highbush Blueberry. Shrub up to 4 m. high, forming compact or open clumps; twigs angled and

warty; buds ovoid or oblong. Swamps, Nova Scotia to Quebec, and Wisconsin, south to Florida and Louisiana (Fig. 271).

6. V. erythrocarpum Michx. Southern Mountain Cranberry. Shrub 3-18 dm. high, deciduous; twigs slender, pubescent, conspicuously angled; buds oblong,appressed, 1-1.5 mm. long. Thickets, in the mountains from Georgia to West Virginia (Fig. 272).

7. V. oxycoccos L. Small Cranberry. Stems trailing, very slender, branches almost capillary; leaves evergreen, 4-10 mm. long, 1-3 mm. broad, strongly revolute, conspicuously whitened beneath; berries present in winter, 6-8 mm. across. Bogs, Greenland to Alaska, south to North Carolina (in the mountains), Ohio, Wisconsin, Minnesota, Saskatchewan, Alberta, and Oregon; also in Eurasia (Fig. 273).

8. V. macrocarpon Ait. Large Cranberry. Stems comparatively thick, elongated; leaves evergreen, 6-18 mm. long, 2-8 mm. broad, pale beneath, flat or slightly revolute; berries present through the winter, 1-2 cm. across. Bogs, Newfoundland to Minnesota, south to North Carolina and Illinois (Fig. 274).

DIOSPYROS L. (Ebenaceae)

Deciduous shrubs or trees. Twigs terete, zigzag, gray-brown; pith moderate, continuous, sometimes becoming porous or chambered. Buds solitary, sessile, deltoid-ovoid, with 2 overlapping scales; terminal bud lacking. Leaf-scars alternate, half-elliptical; bundle-trace 1, curved; stipule-scars none.

1. D. virginiana L. Persimmon. A tree to 30 m. high, usually much smaller, with rounded top and spreading, often pendulous branches; bark dark, deeply divided into square scaly thick plates; fruit a large juicy edible berry, 2-3.5 cm. across, yellowish or pale orange, withering but persistent into winter. Dry woods and old fields, Florida to Texas, north to New England, Pennsylvania, Ohio, Iowa and Kansas (Fig. 275).

HALESIA Ellis (Styracaceae)

Deciduous shrubs or small trees with shredding bark. Twigs moderate, terete; pith small, round, chambered, white. Buds moderate, superposed, ovoid, with about 4 red scales. Leaf-scars alternate, moderate, half-round, notched; bundle-trace 1, curved, compound; stipule-scars none. Fruit a dry capsule, persistent into winter.

1. H. carolina L. Silverbell-Tree. Large shrub or tree up to 10 m. high, with bark separating into small closely appressed scales; twigs glabrate; buds acute, slightly stalked, fruit 2-3.5 cm. long, 4-winged. Rich woods, Florida to Texas, north to Virginia, West Virginia, Illinois, Missouri, and Oklahoma (Fig. 276).

FRAXINUS L. (Oleaceae)

Large deciduous trees. Twigs rather thick, stiff, divergent, sometimes square, or at least compressed at the nodes; pith often obscurely angled. Buds sessile, superposed, with 2 or 3 pairs of scales, those of the terminal bud often lobed. Leaf-scars opposite, half-round to broadly U-shaped; bundle-traces in a curved group; stipule-scars none.

a. Twigs rounded

b. Buds brown or red-brown

c. Leaf-scars deeply notched at the top — 1. F. americana

c. Leaf-scars nearly straight across the top — 2. F. pennsylvanica

b. Buds blue-black — 4. F. nigra

a. Twigs acutely 4-angled — 3. F. quadrangulata

1. F. americana L. White Ash. A tree to 25 m. high, bark gray, furrowed into close diamond-shaped areas; twigs gray, glabrous or glabrate (densely pubescent in the var. biltmoreana (Beadle) J. Wright (F. biltmoreana Beadle), Biltmore Ash); leaf-scars deeply notched at the top; buds dark-brown, the terminal broadly ovoid, obtuse. Rich woods, Quebec to Minnesota, south to Florida and Texas (Fig. 277).

2. F. pennsylvanica Marsh. Red Ash. A tree to 30 m. high, with furrowed bark; twigs gray, velvety pubescent (glabrous in the var. subintegerrima (Vahl) Fernald, (F. lanceolata Borkh.), Green Ash); leaf-scars nearly straight across the top; buds rusty brown, pubescent. River-banks, Georgia and Alabama to Texas, north to New England, Quebec, Michigan, Saskatchewan, and Montana (Fig. 278).

3. F. quadrangulata Michx. Blue Ash. A large tree, usually 20-35 m. high; twigs gray, square, or slightly winged, glabrous; buds gray or reddish-brown; leaf-scars obcordate; bundle-traces

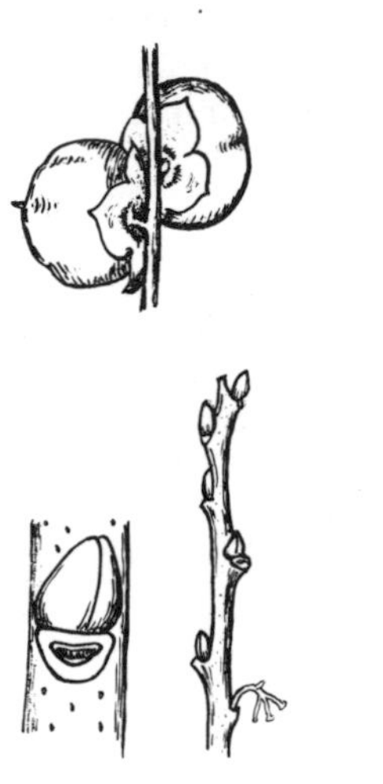

Fig. 275. Diospyros virginiana

Fig. 276. Halesia carolina

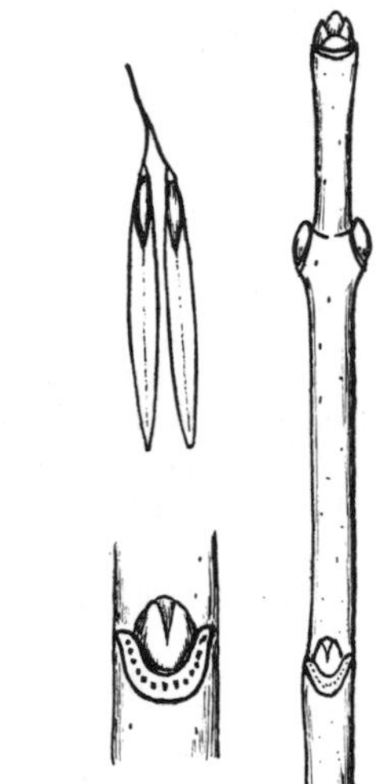

Fig. 277. Fraxinus americana

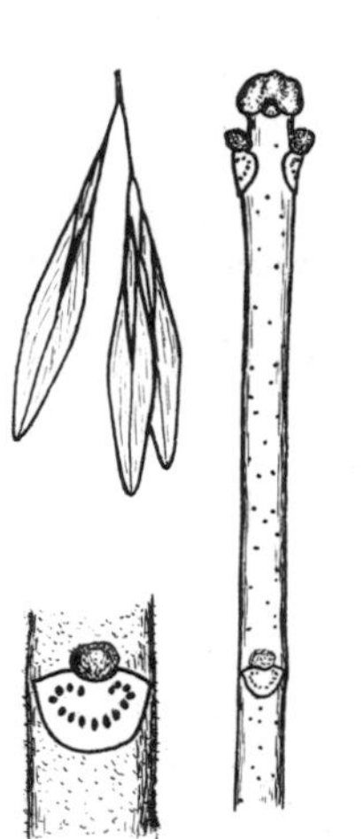

Fig. 278. Fraxinus pennsylvanica

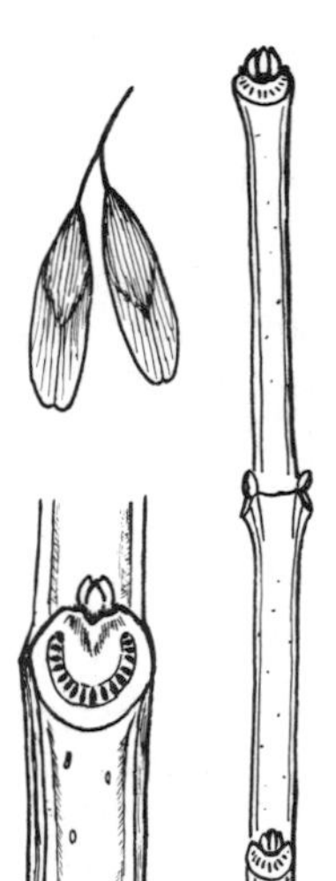

Fig. 279. Fraxinus quadrangulata

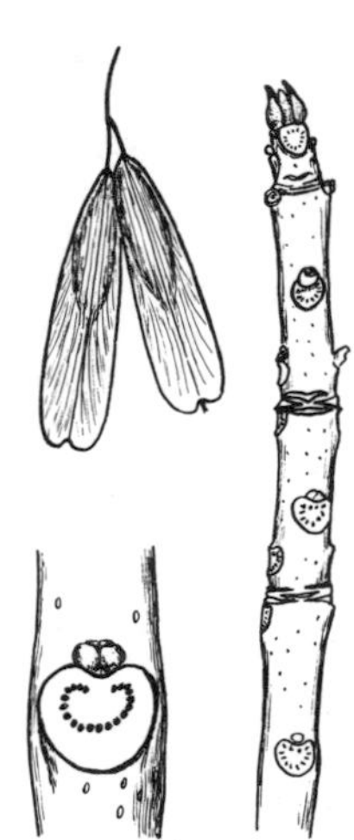

Fig. 280. Fraxinus nigra

in a lunate line. Woods, Ontario, Michigan and Wisconsin, south to Alabama and Oklahoma (Fig. 279).

4. F. nigra Marsh. Black Ash. Tree to 25 m. high, with scaly bark; twigs glabrous; leaf-scars nearly orbicular, the bundle-traces in a nearly closed oval; buds blue-black, the terminal broadly ovate-conical. Swamps, Newfoundland to Manitoba, south to West Virginia, Iowa, and North Dakota (Fig. 280).

SYRINGA L. (Oleaceae)

Upright deciduous shrubs. Twigs moderate, usually somewhat 4-lined; pith moderate, continuous, roundish, pale. Buds solitary (rarely multiple), sessile, ovoid, with about 4 pairs of scales; end-bud often absent. Leaf-scars opposite, small; bundle-trace 1, compound; stipule-scars lacking.

1. S. vulgaris L. Lilac. Upright shrub or small tree to 7 m.; branchlets glabrous; leaf-scars crescent- or shield-shaped, raised; bud-scales fleshy. Introduced from Europe, cultivated as an ornamental, sometimes escaping (Fig. 281).

FORSYTHIA Vahl (Oleaceae)

Loosely branched spreading deciduous shrubs. Twigs somewhat 4-sided, moderate; pith moderate, more or less excavated between the nodes. Buds moderate, becoming multiple, fusiform, sessile, with about 6 pairs of scales. Leaf-scars opposite, shield-shaped, rather small, raised; bundle-trace 1; stipule-scars absent. The showy yellow bell-shaped flowers appear in February or March.

a. Pith solid at the nodes, excavated between the nodes ... 1. F. suspensa
a. Pith all chambered, or in older twigs, all excavated ... 2. F. viridissima

1. F. suspensa (Thunb.) Vahl. Weeping Golden-bells. Shrub to 3 m. high; branchlets pendulous; pith solid at the nodes, but excavated between the nodes. Introduced from Asia, sometimes spreading slightly from cultivation.

2. F. viridissima Lindl. Greenstem Golden-bells. Shrub to 3 m. high; branchlets green, ascending; pith chambered or finally all excavated. Introduced from Asia, spreading slightly from cultivation (Fig. 282).

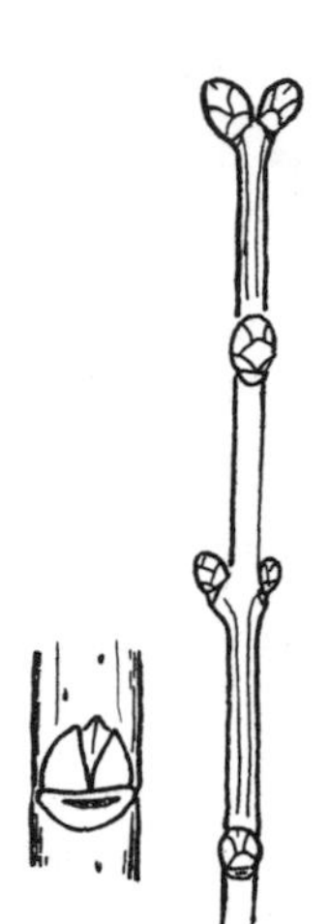

Fig. 281. Syringa vulgaris

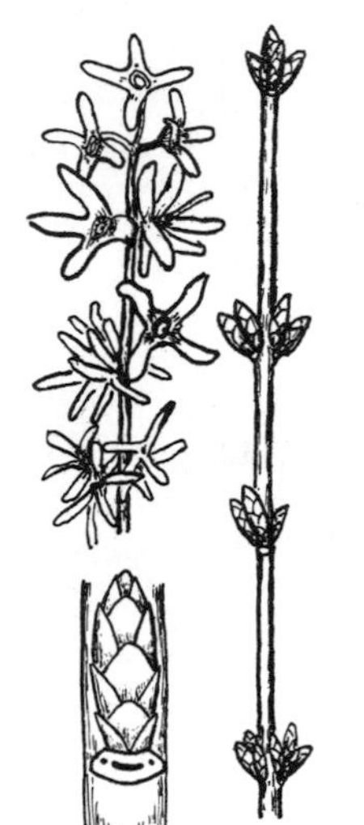

Fig. 282. Forsythia viridissima

Fig. 283. Chionanthus virginicus

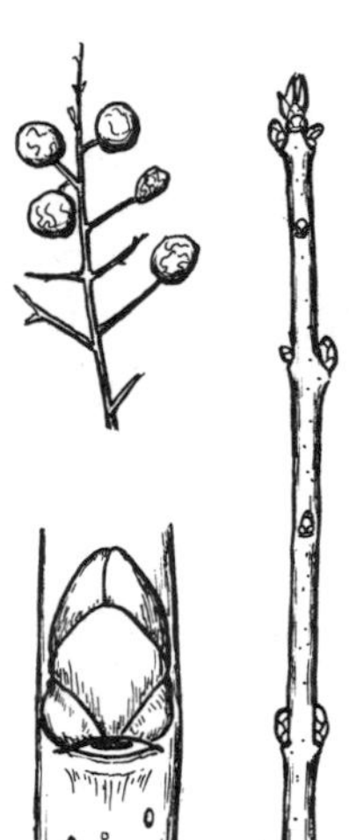

Fig. 284. Ligustrum vulgare

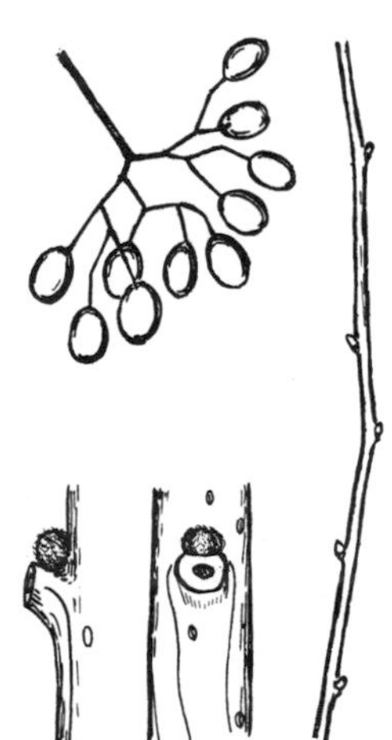

Fig. 285. Solanum dulcamara

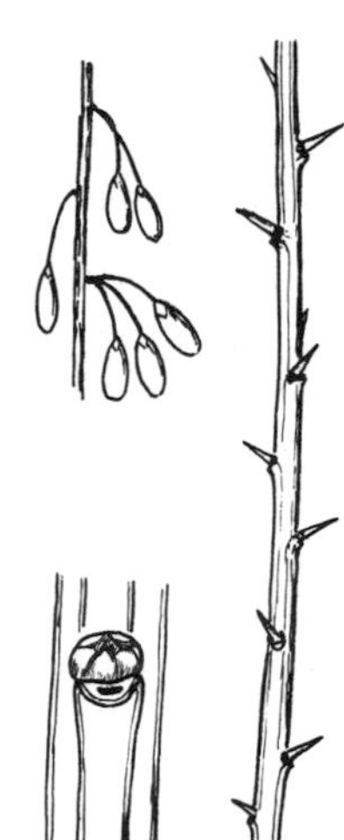

Fig. 286. Lycium halimifolium

CHIONANTHUS L. (Oleaceae)

Deciduous shrubs or small trees. Twigs moderate, somewhat 4-sided; pith moderate, white, continuous. Buds solitary or superposed, sessile, round-ovoid, small, with about 3 pairs of sharp-pointed keeled scales. Leaf-scars opposite, triangular, rather small; bundle-trace 1, compound; stipule-scars none.

1. C. virginicus L. Fringetree. A large shrub or small tree to 20 m. high; twigs moderate, pubescent when young; buds ovoid, buff, the scales sharp-pointed (the points often get broken off during the winter). Damp woods, Florida to Texas, north to New Jersey, West Virginia, Missouri, and Oklahoma (Fig. 283).

LIGUSTRUM L. (Oleaceae)

Deciduous shrubs. Twigs slender, rounded; pith moderate, white, continuous. Buds solitary or superposed, sessile, ovoid, small, with 2 or 3 pairs of scales. Leaf-scars opposite, crescent-shaped or elliptical, small; bundle-trace 1; stipule-scars none.

1. L. vulgare L. Privet. Deciduous shrub (half-evergreen, i.e., holding its foliage into the winter, but dropping it before spring) to 5 m. high; twigs minutely puberulous; bud-scales acute; leaf-scars elliptical. Introduced from Europe, escaped from cultivation and somewhat naturalized (Fig. 284).

SOLANUM L. (Solanaceae)

Deciduous soft-wooded climbers (also herbs). Twigs slender, terete or angled; pith rather large, porous. Buds moderate, solitary, sessile, sub-globose. Leaf-scars alternate, half-round; bundle-trace 1, relatively large; stipule-scars none. Panicle-vestiges with dried berries persistent into winter.

1. S. dulcamara L. Bittersweet. Stems climbing or scrambling, up to 3 m. long, olivaceous, pubescent to glabrate, flexuous, branching. Introduced from Europe, escaped from cultivation and naturalized (Fig. 285).

LYCIUM L. (Solanaceae)

Spreading or scrambling spiny deciduous shrubs. Twigs slender, 5-angled, glabrous, often whitish or with short striations; pith moderate, spongy. Buds small and inconspicuously multiple, subglobose, indistinctly scaly. Leaf-scars alternate, crescent-shaped, small; bundle-trace 1; stipule-scars lacking.

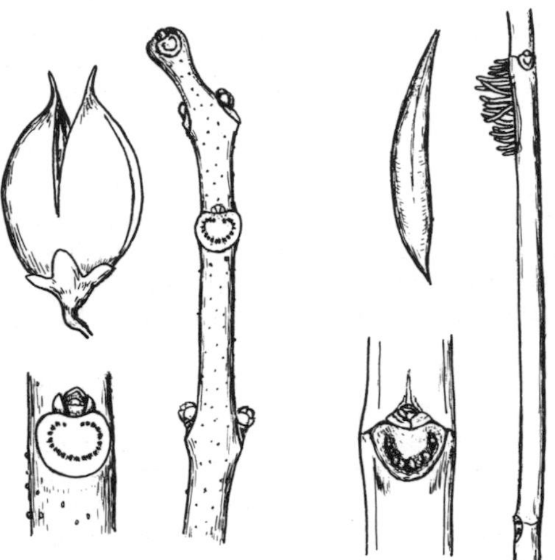

Fig. 287. Paulownia tomentosa Fig. 288. Campsis radicans

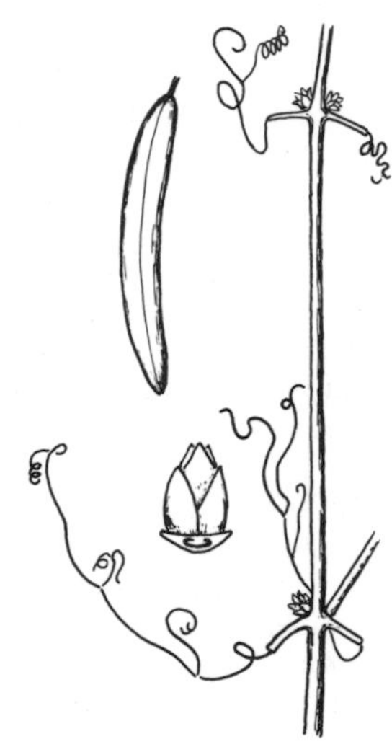

Fig. 289. Bignonia capreolata

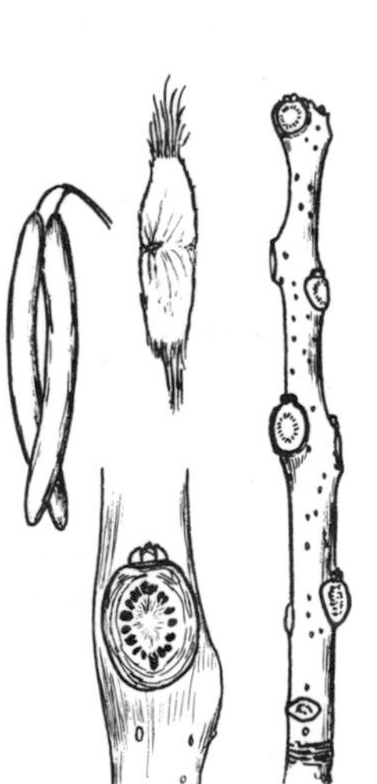

Fig. 290. Catalpa speciosa

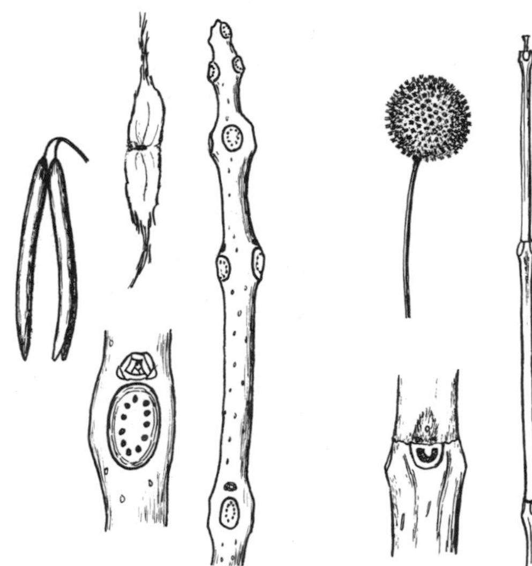

Fig. 292. Cephalanthus occidentalis

Fig. 291. Catalpa bignonioides

1. L. halimifolium Mill. Matrimony-Vine. Shrub with lithe runners, up to 3 m. long, recurved-drooping, sometimes sparingly if at all spiny; twigs pale gray, not hairy. Introduced from Europe and spread from cultivation (Fig. 286).

PAULOWNIA Sieb. and Zucc. (Scrophulariaceae)

Medium-sized deciduous trees. Twigs thick, compressed at the nodes; pith large, white, round, chambered or excavated between the nodes. Buds often superposed, sessile, hemispheric, with about 4 scales; terminal bud lacking. Leaf-scars opposite, large, nearly elliptic; bundle-traces numerous, in a nearly closed ellipse; stipule-scars none. Fruit an ovoid beaked capsule, remaining on the trees into the winter, after the seeds have been shed. By some authorities placed in the Bignoniaceae.

1. P. tomentosa (Thunb.) Steud. Imperial Tree. Tree reaching a height of 15 m. and a diameter of 1 m.; bark thin, flaky; branches thick, spreading; twigs soft-pubescent when young, becoming glabrous; capsules 3-4 cm. long, woody. Introduced from Asia, spread from cultivation and extensively naturalized (Fig. 287).

CAMPSIS Lour. (Bignoniaceae)

Straggling deciduous shrubs, usually climbing, often by aerial roots. Twigs moderate, warty; pith pale, continuous, or excavated at the nodes. Buds small, solitary, sessile, triangular, compressed, with 2 or 3 pairs of visible scales. Leaf-scars opposite, shield-shaped; bundle-trace 1, crescent-shaped, compound; stipule-scars none but leaf-scars connected by hairy ridges. Tecoma Juss.

1. C. radicans (L.) Seem. Trumpet Creeper. Climbing or trailing, to 10 m. or more; aerial roots abundant, in two rows; twigs puberulent or scabrous; capsule elongated, flattened, 8-12 cm. long. Thickets, Florida to Texas, north to New Jersey, West Virginia, and Iowa; naturalized northwards, an aggressive weed southwards (Fig. 288).

BIGNONIA L. (Bignoniaceae)

Soft-wooded deciduous or partly evergreen climbers. Stems rounded or somewhat fluted, slender; pith pale, porous, finally excavated. Buds moderate, solitary, sessile, oblong. Leaf-scars opposite, triangular; bundle-trace 1, U-shaped; stipule-scars none, but leaf-scars connected by transverse ridges. The leaves, composed of 2 leaflets and a branched tendril, sometimes do not disarticulate but persist into the winter.

1. B. capreolata L. Crossvine. Climbing to 20 m. or trailing; twigs glabrous except at the nodes; pith in transverse section as a cross; leaflets stalked, 5-15 cm. long, entire; capsule linear, 10-17 cm. long. Rich woods, Florida to Louisiana, north to Maryland, West Virginia, Illinois, and Missouri (Fig. 289).

CATALPA Scop. (Bignoniaceae)

Small or medium-sized deciduous trees. Twigs thick, round; pith large, round, pale, continuous. Buds solitary, sessile, globose, with about 6 loose scales; terminal bud lacking. Leaf-scars elliptic or round, in whorls, alternately of 2 large scars and 1 small scar, then 1 large scar and 2 small scars; stipule-scars none. Pods long, terete, persistent in winter. Seeds winged all around, the wings ciliate at each end.

a. Capsules about 1.5 cm. thick; hairs of the seeds not coming to a point — 1. C. speciosa
a. Capsules 0.8-1.2 cm. thick; hairs of the seeds coming to a point — 2. C. bignonioides

1. C. speciosa Warder. Catawba-Tree. Cigar-Tree. Northern Catalpa. A large tree to 30 m. tall, of pyramidal habit; bark red-brown, broken into thick scales; capsule thick; seeds truncate. Damp woods, Tennessee to Texas, north to Indiana and Iowa; cultivated and naturalized elsewhere (Fig. 290).

2. C. bignonioides Walt. Common Catalpa. A low tree to 15 m. high, or taller, with wide-spreading branches forming a broad round head; bark light brown, separating into thin scales; pods slender; seeds pointed. Native of the southern and Gulf states, cultivated and frequently escaped northwards (Fig. 291).

CEPHALANTHUS L. (Rubiaceae)

Deciduous shrubs. Twigs slender, round, glabrous; pith small, more or less angled, brown, continuous. Buds solitary, sessile, conical, in depressed areas above the leaf-scars; terminal bud lacking. Leaf-scars in whorls of 3, or opposite, roundish; bundle-trace 1, U-shaped; stipule-scars or persistent stipules connecting the leaf-scars.

1. C. occidentalis L. Buttonbush. Deciduous shrub to 5 m. tall; twigs reddish and glossy. Swamps and stream-margins,

Florida to Mexico, north to Nova Scotia, Ontario, Oklahoma, and California; also in the West Indies (Fig. 292).

DIERVILLA Duham. (Caprifoliaceae)

Deciduous shrubs. Twigs terete, moderate, straw-colored or brownish; pith pale, continuous, moderate. Buds solitary or superposed, sessile, oblong, appressed. Leaf-scars opposite or occasionally whorled, triangular, moderate, connected by lines; bundle-traces 3; stipule-traces none. Capsules linear, persisting in winter.

1. D. lonicera Mill. Bush Honeysuckle. Upright soft-wooded shrub 1 m. or less high; bark grayish-brown; twigs with two ridges decurrent from the nodes, glabrous except for the ridges; capsule glabrous, 8 mm. long, beaked. Dry rocky places, Newfoundland to Manitoba, south to North Carolina, Indiana, and Iowa (Fig. 293).

LONICERA L. (Caprifoliaceae)

Deciduous or partly evergreen shrubs or woody climbers. Twigs rounded, slender; pith moderate, continuous, or excavated between the nodes. Buds solitary or superposed, sessile. Leaf-scars opposite, crescent-shaped, small, on the narrowed tips of leaf-bases, more or less connected by lines; bundle-traces 3; stipule-scars none.

a. Erect shrubs; buds often superposed

- b. Pith brown, excavated between the nodes — 1. L. tatarica
- b. Pith white, continuous — 2. L. canadensis

a. Twining or loosely ascending shrubs

- b. Stems red-brown, hairy, nearly evergreen — 3. L. japonica
- b. Stems gray or straw-colored
  - c. More or less evergreen, at least southwards — 4. L. sempervirens
  - c. Deciduous — 5. L. dioica

1. L. tatarica L. Tartarian Honeysuckle. Erect smooth shrub 1.5-3 m. high; branches slender, glabrous; pith brown, excavated between the nodes; buds glabrous, oblong or ovoid, the scales short-pointed. Introduced from Eurasia, frequently escap-

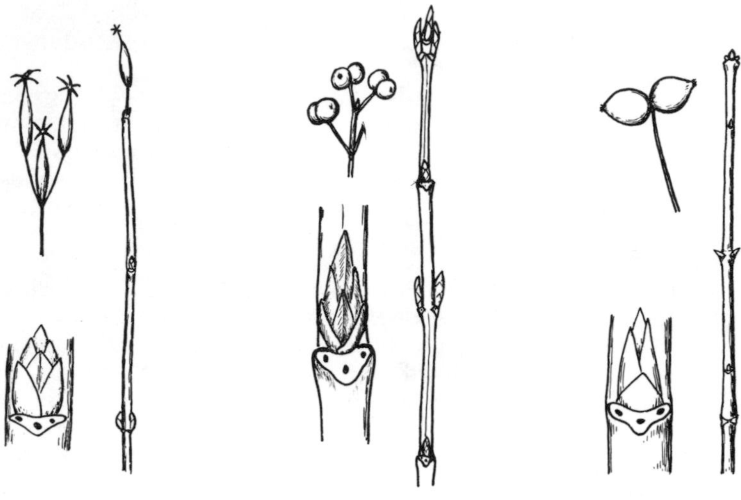

Fig. 293. Diervilla lonicera

Fig. 294. Lonicera tatarica

Fig. 295. Lonicera canadensis

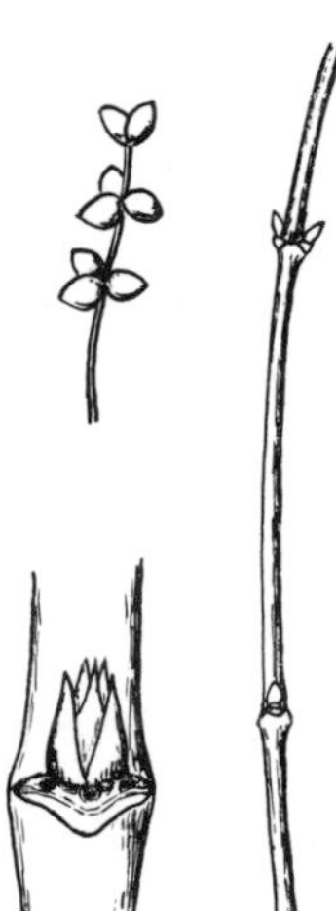

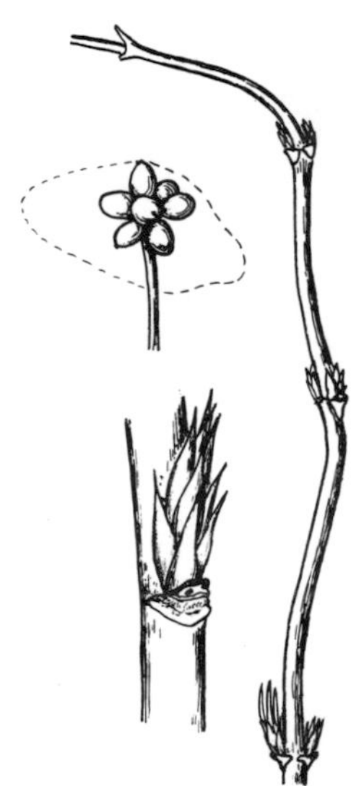

Fig. 296. Lonicera japonica

Fig. 297. Lonicera sempervirens

Fig. 298. Lonicera dioica

ing from cultivation (Fig. 294).

2. L. canadensis Bartr. American Fly Honeysuckle. A straggling shrub 1-1.5 m. high; twigs glabrous; buds short-ovoid or nearly globose, glabrate, scales acute, the lower distinctly shorter than the bud. Cool woods, Quebec to Saskatchewan, south to North Carolina, Indiana and Iowa (Fig. 295).

3. L. japonica Thunb. Japanese Honeysuckle. A half-evergreen twining or trailing shrub with pubescent red-brown twigs; leaves ovate, 2-8 cm. long, entire. Introduced from Asia, much planted, especially along roadsides, now widely spreading and naturalized, often a very aggressive weed (Fig. 296).

4. L. sempervirens L. Trumpet Honeysuckle. High-climbing glabrous shrub, more or less evergreen; twigs gray or straw-colored; leaves elliptic, 3-8 cm. long, glaucous. Woods and thickets, Florida to Texas, north to Maine, Ohio, Iowa, and Nebraska, partly as an escape from cultivation northwards (Fig. 297).

5. L. dioica L. Smooth Honeysuckle. A slightly climbing or erect deciduous shrub, 1-3 m. high; bark grayish, peeling on old stems; twigs glaucous and glabrous; buds ovoid, the scales ovate, the lowest about as long as the buds. Rocky banks, Maine and Quebec to Manitoba and British Columbia, south to Georgia, Missouri, and Kansas (Fig. 298).

SYMPHORICARPOS Duham. (Caprifoliaceae)

Erect much branched deciduous shrubs. Twigs slender, round, pubescent; pith small, round, brownish, continuous or excavated. Buds small, solitary or multiple, ovoid-oblong, sessile. Leaf-scars opposite, half-round, small and torn, partly connected by ridges; bundle-trace 1, indistinct; stipule-scars none. Fruit a berry persistent into winter.

a. Pith excavated; fruit white 1. S. albus
a. Pith continuous; fruit red 2. S. orbiculatus

1. S. albus (L.) Blake. Snowberry. (S. racemosus Michx.) Erect shrub 1-1.5 m. high; twigs glabrate; buds 2 mm. long; fruit globose, white, 5-10 mm. in diameter when fresh, withering but persisting into winter. Rocky limestone slopes, Quebec to British Columbia, south to Ohio, Michigan, Minnesota, Nebraska, and Colorado (Fig. 299).

2. S. orbiculatus Moench. Coralberry. Shrub to 2 m. high, with slender upright pubescent branches; fruit subglobose, red,

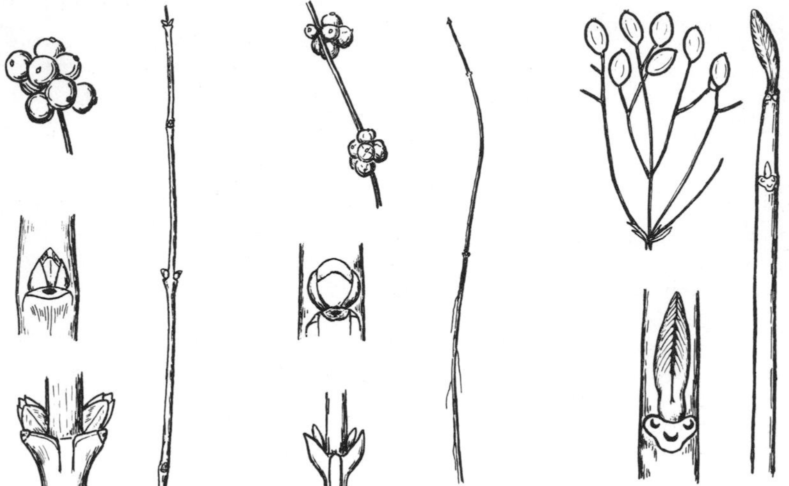

Fig. 299. Symphoricarpos albus

Fig. 300. Symphoricarpos orbiculatus

Fig. 301. Viburnum alnifolium

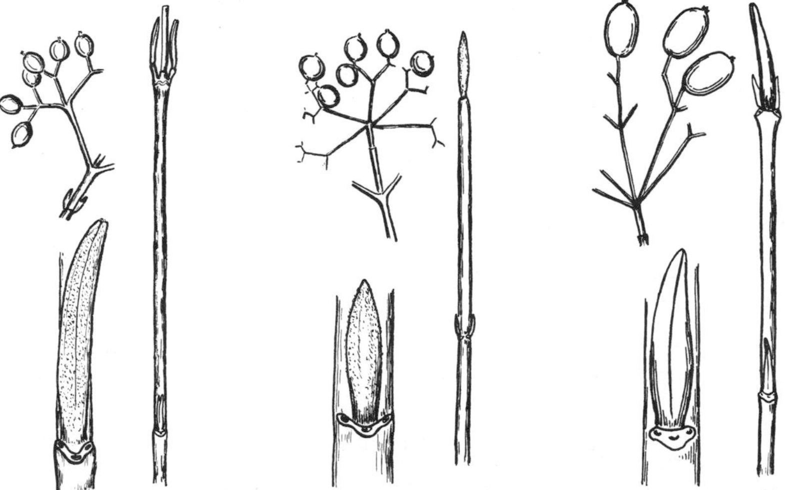

Fig. 302. Viburnum cassinoides

Fig. 303. Viburnum nudum

Fig. 304. Viburnum lentago

4-6 mm. across when fresh, withering but persisting into winter. Thickets, Florida to Texas and Mexico, north to Pennsylvania, Illinois, Minnesota, South Dakota, and Colorado; spread from cultivation northeastwards (Fig. 300).

VIBURNUM L. (Caprifoliaceae)

Deciduous shrubs or small trees. Twigs moderate, slender, obscurely 6-sided; pith moderate, continuous. Buds solitary or superposed, mostly stalked, ovoid or oblong. Leaf-scars opposite, crescent-shaped or broad, often meeting or connected by lines; bundle-traces 3; stipule-scars none. Drupes present in winter.

a. Leaf-scars quite broad; twigs purple, stellate-scurfy — 1. V. alnifolium
a. Leaf-scars narrow; twigs brown or gray

  b. Bud-scales 2, closely valvate

    c. Buds ovoid-globose, green — 10. V. trilobum
    c. Buds oblong, brown-scurfy or lead-colored

      d. Branches numerous, often short and rigidly spreading — 5. V. prunifolium
      d. Branches elongate, fewer, flexuous

        e. Buds smooth, lead-colored — 4. V. lentago
        e. Buds brown, scurfy

          f. Twigs dull — 2. V. cassinoides
          f. Twigs glossy — 3. V. nudum

  b. Bud-scales more than 2, the lower pair mostly short

    c. Twigs not stellate-pubescent

      d. Bud-scales 4, buds appressed

        e. Lower bud-scales very short; twigs pubescent — 9. V. acerifolium
        e. Lower scales often half as long as the bud; twigs glabrous — 8. V. recognitum

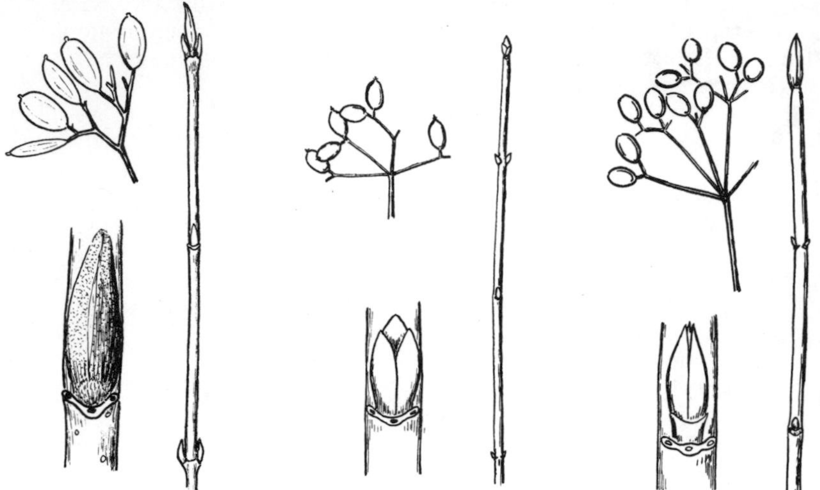

Fig. 305. Viburnum prunifolium

Fig. 306. Viburnum rafinesquianum

Fig. 307. Viburnum dentatum

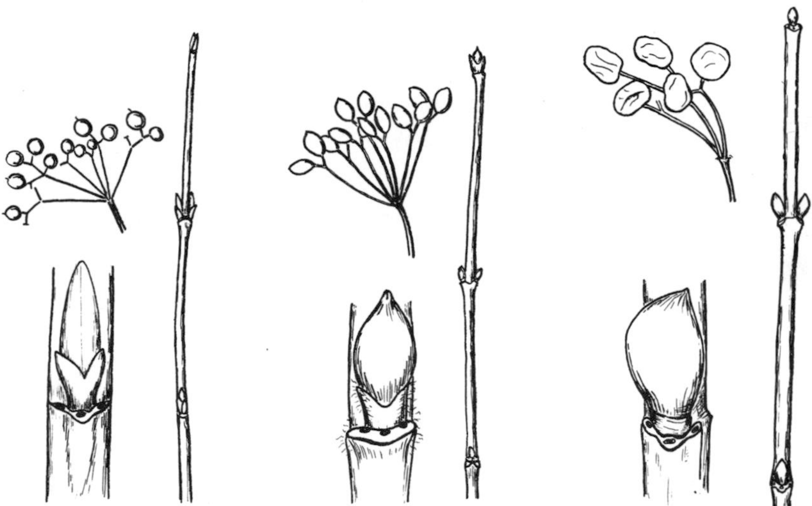

Fig. 308. Viburnum recognitum

Fig. 309. Viburnum acerifolium

Fig. 310. Viburnum trilobum

d. Bud-scales often 6, buds divergent, plump — 6. V. rafinesquianum

c. Young twigs densely stellate-pubescent, becoming glabrate — 7. V. dentatum

1. V. alnifolium Marsh. Hobblebush. Straggling shrub to 3 m. high, with forked branches, often procumbent and rooting, so as to trip pedestrians (whence the common name); twigs scurfy-pubescent. Cool woods, Prince Edward Island to Ontario, south in the mountains to Georgia and Tennessee (Fig. 301).

2. V. cassinoides L. Wild Raisin. Upright shrub 1-4 m. high, or sometimes taller; twigs dull, scurfy, elongated, flexuous; buds covered by a single pair of yellow or golden scurfy scales; drupes blue-black, 6-9 mm. long. Thickets, Newfoundland to Ontario, south to Wisconsin, Indiana, and in the mountains to Alabama (Fig. 302).

3. V. nudum L. Smooth Witherod. Swamp-Haw. Upright shrub or small tree to 6 m. high, and 1-2 dm. in diameter, with slightly scurfy, rather glossy, elongated, flexuous twigs; buds brown or fuscous. Swamps, Florida to Texas, north to Connecticut and Kentucky; mostly in the coastal plain or Mississippi Valley (Fig. 303).

4. V. lentago L. Sheepberry. Nannyberry. A shrub or small tree to 10 m. high, with slender branches, slightly scurfy twigs, and gray buds, the terminal long-pointed; drupes blue-black, 0.8-1.5 cm. long, with sweet pulp. Stream-banks, Quebec to Manitoba and South Dakota, south to Georgia, Missouri, and Colorado (Fig. 304).

5. V. prunifolium L. Black Haw. Large shrub or small tree to 8 m. high; bark blackish, broken into squarish blocks; branches numerous, rigid, spreading; buds short-pointed, reddish, pubescent; twigs glabrous; drupes blue-black, about 1 cm. long. Thickets, Florida to Texas, north to Connecticut, Michigan, Iowa, and Kansas (Fig. 305).

6. V. rafinesquianum Schultes. Downy Arrowwood. (V. pubescens of authors, not (Ait.) Pursh). Loose straggling or dense shrub up to 2 m. high; buds with 2 pairs of outer scales; branchlets glabrous, pale; fruit dark purple, ellipsoid, 7-9 mm. broad. Dry slopes, Quebec to Manitoba, south to Georgia, Kentucky, and Arkansas (Fig. 306).

7. V. dentatum L. Southern Arrowwood. (V. scabrellum (T. and G.) Chapm.). Shrub 1-3 m. high, with close gray bark;

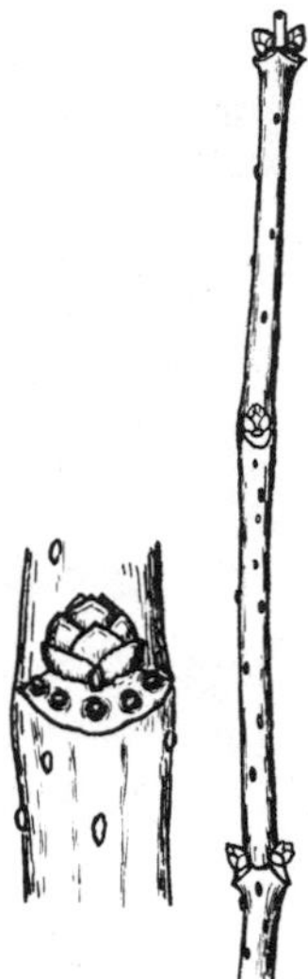

Fig. 311. Sambucus canadensis

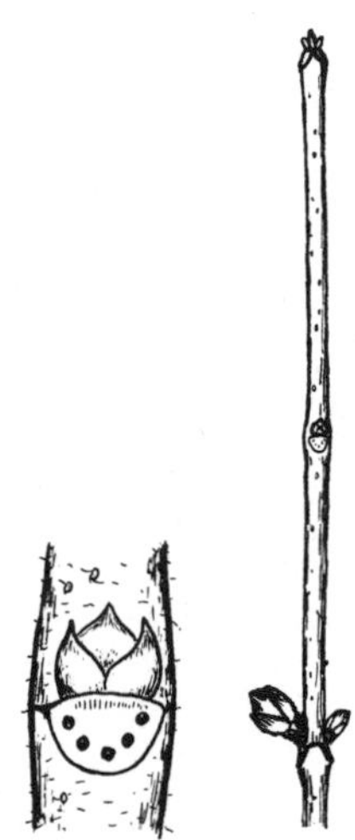

Fig. 312. Sambucus pubens

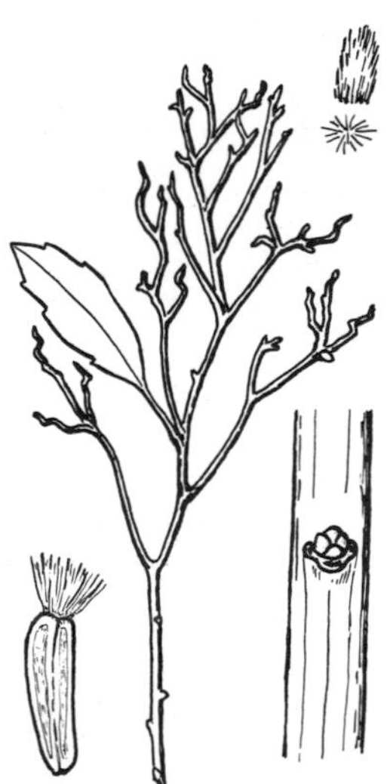

Fig. 313. Baccharis halimifolia

young branchlets often densely pubescent, sometimes glabrate; drupes blue-black, ellipsoid-ovoid, 5-10 mm. long. Sandy thickets, Florida to Texas, north to Massachusetts, West Virginia, Tennessee, and Missouri (Fig. 307).

8. V. recognitum Fernald. Smooth Arrowwood. (V. dentatum of authors, not L.). Shrub 1-3 m. high; branchlets glabrous. Damp thickets, New Brunswick to Ontario, south to South Carolina, Ohio, and Michigan (Fig. 308).

9. V. acerifolium L. Mapleleaf Arrowwood. Dockmackie. Upright shrub to 2 m. high, with pubescent twigs; buds stalked, appressed, with 4 scales; drupes ellipsoid, purple-black. Rocky woods, Quebec to Minnesota, south to Georgia and Tennessee (Fig. 309).

10. V. trilobum Marsh. Cranberry-tree. (V. opulus L. var. americanum Ait.). A nearly smooth upright shrub, 1-4 m. high, with gray bark; twigs glabrous; buds with 2 connate outer scales; drupe orange to red, subglobose to ellipsoid, juicy. Cool woods, Newfoundland to British Columbia, south to West Virginia, Indiana, Iowa, South Dakota, Wyoming, and Washington (Fig. 310).

SAMBUCUS L. (Caprifoliaceae)

Deciduous soft-wooded shrubs or exceptionally small trees. Twigs more or less angled, thick; lenticels very conspicuous; pith very large, soft, continuous. Buds solitary or multiple; terminal bud mostly lacking. Leaf-scars opposite, broadly crescent-shaped or triangular, large, more or less connected by lines.

a. Pith white	1. S. canadensis
a. Pith brown	2. S. pubens

1. S. canadensis L. Common Elderberry. Black Elderberry. A stoloniferous shrub to 4 m. high, with pale yellowish-gray twigs. Damp rich soil, Nova Scotia to Manitoba, south to Georgia, Louisiana, and Oklahoma (Fig. 311).

2. S. pubens Michx. Red Elderberry. (S. racemosa of authors, not L.). Shrub to 4 m. tall, with light brown twigs. Rocky woods, Newfoundland to Alaska, south to Georgia, Ohio, Iowa, South Dakota, Colorado, and Oregon (Fig. 312).

BACCHARIS L. (Compositae)

Soft-wooded tardily deciduous shrubs with resin-passages. Twigs slender, about 8-ridged, green; pith small, crenulate, pale, continuous. Buds rather small, solitary, sessile, nearly globose, with about 4 exposed scales, more or less enclosed in hardened greenish resin. Leaf-scars alternate, small, angularly crescent-shaped, slightly raised, decurrent in ridges from the angles; bundle-traces 3; stipule-scars lacking.

1. B. halimifolia L. Groundsel-Tree. Shrub 1-3 m. high, glabrous but somewhat scurfy; branches angled. Thickets, on the coastal plain, Florida to Texas and Mexico, north to Massachusetts (Fig. 313).

BIBLIOGRAPHY

Bailey, L.H. The Standard Cyclopedia of Horticulture. The Macmillan Co. New York, 1935.

Blakeslee, A.F. and C.D. Jarvis. New England trees in winter. Bull. No. 69, Storrs Agri. Exp. Sta. pp. 307-576. 1911.

Bode, I.T. and G.B. MacDonald. A Handbook of the Native Trees of Iowa. Iowa State College, Ames. 1914.

Brooks, A.B. West Virginia Trees. W. Va. Agri. Exp. Sta. Bull. 175. 1920.

Core, Earl L. Plant Taxonomy. Prentice-Hall. Englewood Cliffs, N.J. 1955.

Core, Earl L. and Nelle Ammons. Woody Plants of West Virginia in Winter Condition. Edwards Bros., Ann Arbor, Mich. 1946.

Fernald, M.L. Gray's Manual of Botany, 8th ed. American Book Co. New York. 1950.

Gleason, Henry A. The new Britton and Brown illustrated flora of the northeastern United States and adjacent Canada. 3 vols. The New York Botanical Garden. New York. 1952.

Harlow, William M. Twig Key to the Deciduous Woody Plants of Eastern North America. Published by the author. Syracuse, N.Y. 4th ed. 1941.

Harlow, William M. and Ellwood S. Harrar. Textbook of Dendrology. McGraw-Hill. 4th ed. 1958.

Hough, R.B. Handbook of the Trees of the Northern States and Canada. The Macmillan Co., 1955.

Illick, Joseph S. Pennsylvania Trees. Penna. Dept. For. Waters, Bull. 11. Harrisburg. 1928.

Muenscher, W.C. Keys to Woody Plants. 6th ed. Ithaca. 1950.

Otis, Charles Herbert. Michigan Trees. Univ. of Mich. Bull. N.S., XXII, No. 26. 1920.

Rehder, Alfred. Manual of Cultivated Trees and Shrubs. Macmillan. 2nd ed. 1940.

Sargent, F. L. Key to Common Deciduous Trees in Winter and Key to Common Woods. Cambridge, Mass. 1903.

Strausbaugh, P. D. and Earl L. Core. Trees and Shrubs of West Virginia. Mimeographed. Morgantown. 1935.

Trelease, William. Winter Botany. Published by the author. Urbana, Ill. 1918.

Wiegand, K. M. and F. W. Foxworthy. A key to the genera of woody plants in winter. Ithaca, N. Y. 1904. 2nd ed., 1906. 3rd. ed. 1908.

GLOSSARY

Acuminate. Gradually tapering to a long point.
Acute. Tapering to a point, but not long-pointed; terminating with a well-defined or sharp angle.
Adnate. (Of stipules) Attached wholly or in part to the base of the petiole.
Alternate. (of leaves) Occurring one at each node, scattered singly along the stem.
Annular. In the form of a ring; said of leaf scars which encircle the bud, or of bundle scars which are circular with an opening in the center.
Apiculate. Ending in a short, pointed tip.
Apophysis. The exposed portion of the cone scale of a conifer.
Appressed. (Of pubescence) Lying close or flat against the blade or petiole; (of buds) lying flat against the stem.
Arachnoid. Cobweb-like.
Arborescent. Tree-like in appearance, size and growth.
Armed. Bearing thorns, spines, or prickles.
Aromatic. Fragrant, having an agreeable smell or taste.
Ascending. Arising somewhat obliquely, or curving upwards.
Auriculate. With basal lobes.
Awl-shaped. Tapering from the base to a slender or rigid point.
Awn. A long stiff hair or hair-like point.
Axil. The upper angle formed by a leaf or branch with the stem.
Axillary. Situated in an axil.
Axis. The central line of support, as a stem.
Berry. A juicy or fleshy fruit in which the seeds are embedded in the pulp.
Biennial. A plant in which growth begins in the spring, summer, or autumn of one growing season, and flowering and death occur the following year.
Bipinnate. Twice-pinnate.
Bipinnatifid. Twice-pinnatifid.
Biternate. Twice-ternate.
Blade. The flat, expanded part of a leaf.
Bloom. A powdery or somewhat waxy substance easily rubbed off.
Bract. A small leaf-life appendage below a flower or a flower cluster.
Bud. A rudimentary branch, during winter in the resting stage. Scaly buds are protected by modified leaves or stipules (scales), while Naked buds lack this protection.
Bulb. An underground bud with fleshy scales and a short axis.
Bundle scars. Small dots or lines on the surface of the leaf scar; they are the scars of the conducting strands that served the leaf.
Bundle traces. Same as bundle scars.

Callose. With a small, hard protuberance.
Calyx. The outer portion of a flower which sometimes persists.
Caespitose. Growing in compact tufts.
Cane. The long shoots of certain shrubs, as the raspberries, blackberries, etc.
Capsule. A dry fruit of two or more carpels, usually splitting into valves at maturity.
Carpel. The part of the flower or fruit which actually bears the seeds.
Catkin. An elongated scaly cluster of flowers.
Chambered. Said of pith when divided into small compartments separated by transverse partitions.
Channelled. Deeply grooved longitudinally.
Ciliate. With marginal hairs, especially if the hairs are in definite lines.
Ciliolate. Minutely ciliate, but visible with the unaided eye.
Clasping. (Of the base of a petiole) Partly or wholly surrounding the stem.
Cleft. Lobed with the sinuses extending about half way to the midvein.
Collateral. Said of extra buds which occur on either side of an axillary bud.
Cone. The reproductive structure of gymnosperms, consisting of an axis to which are attached many woody, overlapping scales which bear seeds.
Confluent. (Of leaflets) More or less united along the midvein.
Connate. (Of leaves) More or less united at the bases of the petioles.
Continuous. Said of pith which is solid, not interrupted by cavities.
Compound leaf. A leaf in which the blade consists of two or more separate leaflets.
Cordate. (Of the base of a leaf) With two, usually rounded, lobes and a sinus; (of a leaf) heart-shaped.
Coriaceous. Leathery in texture.
Corm. The enlarged, fleshy, solid base of a stem (underground).
Corymb. A flat-topped or convex flower cluster, the outer flowers opening first.
Cotyledons. The first leaf or pair of leaves in the embryo which may be expanded and persistent in certain biennials.
Creeping. Running along at or near the surface of the ground and rooting.
Crenate. With rounded teeth; scallop-toothed.
Crenulate. Finely crenate, the teeth small and shallow.
Crisped. (Of margins) Puckered, ruffled.
Cuneate. (Of the base of a blade) Like the acute angle of a wedge, the narrow end pointing toward the base; (of a blade or leaf) triangular, with the acute angle downward.

Deciduous. Falling off, usually at the end of the season.
Decompound. More than once-compound, the primary divisions divided at least once again.
Decumbent. Stems or branches reclining, but the ends ascending.
Decurrent. Continued down the twig in a ridge or wing, as applied to leaf-bases.
Deltoid. Shaped like the Greek letter Δ.
Dentate. Toothed, with the teeth pointing outward.
Denticulate. Finely dentate, the teeth small and shallow.
Depressed. Somewhat flattened from above.
Diaphragmed. Said of pith which is solid with transverse bars of denser tissue at short intervals.
Digitately compound. Leaflets diverging, like the fingers spread.
Dilated. (Of the bases of a petiole) Conspicuously broadened, the broadened part usually much thinner, at least near the margins.
Dissected. Cut or divided into numerous narrow segments.
Distinct. Separate; (of leaf segments or leaflets) not confluent with other segments or leaflets by a winged rachis or midvein.
Divergent. Extending out. Said of buds which point away from the twigs.
Divided. Cleft to the base or to the midvein.
Ellipsoid. An elliptical solid.
Elliptic. With the outline of an ellipse.
Emarginate. With a notch at the tip.
Entire. Without divisions, lobes, or teeth.
Epidermis. The outer layer or covering of plants.
Evergreen. Not falling at end of growing season, having green leaves in winter.
Excavated. Hollowed out; hollow.
Exfoliating. Peeling away.
Expanded. (Of the base of a petiole) Conspicuously broadened, the broadened part not becoming very thin (cf. dilated).
Fascicle. A close cluster or bundle, as of Pine leaves.
Fibrous. Resembling fibers.
Filiform. Thread-shaped; long, slender, and circular in cross-section.
Fimbriate. Fringed.
Flaccid. Without rigidity.
Flexuous. Zigzag; bending alternately in opposite directions.
Fluted. Grooved longitudinally.
Foliaceous. Leaf-like in appearance.
Foliolate. With separate leaflets.
Forked. Divided into nearly equal branches.
Fusiform. Spindle-shaped; swollen in the middle and tapering toward each end.
Glabrate. Almost without hairs; with occasional hairs.

Glabrescent. Becoming glabrous.
Glabrous. Without hairs.
Gland. A small protuberance, consisting of one or more secreting cells.
Glandular. Bearing glands, or any protuberance having the appearance of a gland.
Glaucous. With a bluish-white bloom which may be rubbed off.
Globose. Spherical or nearly so, globular.
Granular. Minutely roughened.
Hastate. Shaped like an arrowhead, but with the basal lobes pointing outward nearly at right angles.
Heath-like. Fine-stemmed and low, with small persistent leaves.
Herbaceous. Herb-like, soft.
Hirsute. Pubescent with rather coarse or stiff, usually relatively long hairs.
Hispid. Pubescent with bristly, rigid hairs.
Hoary. Covered with close, whitish or grayish-white hairs.
Imbricated. Overlapping, as the shingles of a house.
Incised. Cut sharply and more or less deeply and irregularly into lobes.
Inequilateral. Unequal-sided; oblique at the base.
Inflorescence. The flowering portion of a plant, and especially its arrangement.
Introduced. Brought in from another region, especially from Europe or Asia, intentionally or accidentally.
Keel. A ridge on the back of a bud-scale, etc.
Key. A fruit furnished with a wing or leaf-like expansion.
Lacerate. Irregularly cleft as if torn.
Laciniate. Cut into narrow, pointed lobes or segments.
Lanceolate. About four times longer than broad, widest about a third above the base.
Lateral. Said of buds which occur on the sides of the twig.
Leaf-base. See leaf-cushion.
Leaf-cushion. A raised base on which the leaf-scar sometimes appears.
Leaflet. One of the divisions of a compound leaf.
Leaf-scar. A patch differing in color and texture from the rest of the twig and representing the place from which the leaf has fallen.
Legume. A simple dry fruit, opening along two lines, as in beans.
Lenticels. Small areas of loose tissue which appear as dots or warts on the surface of twigs.
Linear. Long and narrow, with nearly parallel margins, at least six times longer than broad.
Lobed. A rounded segment or division of any organ; (of a blade) more or less cut toward the midvein or base.

Longitudinally-veined. With the principal veins parallel or nearly so to the apex, or almost to the apex.
Lyrate. Pinnatifid, with a large, broad, rounded terminal lobe and small basal lobes.
Marcescent. Withering without falling off.
Membranous. Thin, rather soft, and more or less translucent.
Mucronate. With a short, sharp, abrupt tip.
Mucronulate. Mucronate, but the tip very small.
Multiple buds. Several buds in or over an axil, instead of the customary solitary bud.
Naked. Said of a bud which is not covered by scales.
Node. The region of a stem from which one or more leaves arise.
Oblanceolate. Lanceolate, but broadest about a third below the apex.
Oblique. Unequal-sided or slanting.
Oblong. Longer than broad, with the sides nearly parallel.
Obovate. Ovate, but broadest above the middle.
Obtuse. Blunt or rounded.
Olivaceous. Olive-green.
Opaque. Not shining or transparent.

Orbicular. Approximately circular in outline.
Oval. Broadly elliptic, about 1-1/2 times longer than broad.
Ovate. In outline like a longitudinal section of a hen's egg, broadest below the middle.
Ovoid. Egg-shaped, with the broadest portion near the base.
Palmately-veined. With the principal veins arising from the same point at the base of the blade.
Panicle. A branched cluster of flowers.
Papillose. With minute, blunt projections on the surface.
Parted. Deeply cleft; cleft nearly but not quite to the base.
Pedicel. The stalk of a single flower.
Peduncle. The stalk of a flower cluster.
Perennial. A plant living, and usually reproducing, through more than two growing seasons.
Persistent. Remaining after flowering, fruiting, or maturing.
Petiole. The unexpanded part of a leaf.
Pilose. Having long, soft hairs.
Pinnately-compound. With the blade divided into distinct leaflets or segments along a common axis.
Pinnately-veined. With the lateral veins arranged along the two sides of the midvein, not arising from a single point.
Pinnatifid. Pinnately cleft to the middle or beyond.
Pith. The soft, spongy tissue in the center of stems and branches.
Plicate. Folded into plaits.
Polster. A cushion-like mass of vegetation.

Pome. A fleshy fruit of which the apple is a typical example.
Prickle. A spine-like outgrowth of the epidermis.
Prickly. With prickles.
Prostrate. Lying flat on the ground.
Puberulent. With very short hairs; minutely pubescent.
Pubescent. With hairs.
Punctate. With translucent or colored dots or depressions.
Pungent. Sharp pointed.
Pyriform. Pear-shaped.
Raceme. An inflorescence with flowers borne on pedicels of equal length, and arranged on an elongated axis.
Rachis. The axis of a compound leaf.
Rank. A row; 2-ranked meaning in 2 rows.
Recurved. Curved backward or downward.
Reflexed. Bent sharply backward.
Reniform. Kidney-shaped and broader than long.
Repand. With a slightly uneven and somewhat sinuate margin.
Resinous. Having resin.
Reticulate. Arranged as in a network.
Retrorse. Turned backward or downward.
Retuse. With a shallow notch at a rounded apex.
Revolute. Rolled backward from the edge.
Rhizome. A horizontal underground stem, usually rooting at the nodes and becoming erect at the apex.
Rosette. A cluster of prostrate leaves having a radially symmetrical arrangement.
Rostrate. Bearing a beak.
Rugose. Wrinkled.
Runcinate. Sharply pinnatifid or incised, the lobes or segments turned backward.
Sagittate. Shaped like an arrowhead, the basal lobes pointing downward.
Samara. A winged fruit, a key.
Scabrous. Rough to the touch when rubbed in at least one direction.
Scale. A small, rudimentary leaf.
Scarious. Thin, dry, and translucent, not green.
Scurfy. Covered with small bran-like scales.
Segment. One of the parts of a cleft or divided blade.
Serrate. With sharp teeth pointing forward.
Serrulate. Finely serrate, the teeth small and shallow.
Sessile. Without a petiole or stalk.
Seta. A bristle.
Sheath. The tubular lower part of a leaf enclosing the stem.
Sheathing. Enclosing as by a sheath.
Shrub. A low woody growth which usually branches near the base; a bush.
Sinuate. With a strongly wavy margin.

Sinus. The space between two lobes of a blade.
Spatulate. Spatula-shaped or spoon-shaped.
Spine. A sharp outgrowth from the stem.
Spinulose. With small, sharp spines.
Spreading. (Of pubescence) The hairs erect, suberect, or ascending.
Spur-shoots. Short stubby branches with greatly crowded leaf scars and very slow growth.
Stalk. The unexpanded, basal part of a leaflet.
Stellate. (Of pubescence) With three or more radiating branches from the ends of the hairs.
Sterigmata. Very small leaf stalks of certain conifers.
Stipule. One of a pair of appendages at the base of a petiole, often adnate to it.
Stipule-scar. Small marks or lines left by deciduous stipules.
Stolon. A runner, or any basal branch that roots at the nodes.
Stoloniferous. With stolons.
Striate. Marked with fine longitudinal lines.
Strigose. With appressed or ascending stiff hairs.
Subacuminate. Somewhat acuminate.
Subacute. Acutish, somewhat acute.
Suborbicular. Nearly orbicular.
Subpinnatifid. Nearly pinnatifid.
Subulate. Awl-shaped.
Succulent. Juicy.
Superposed. Said of extra buds which appear above the true axillary buds; usually flower buds.
Tapering. Gradually becoming smaller in diameter or width toward one end.
Tendril. A long, slender, coiling structure serving as the organ of attachment in some climbing plants.
Terete. Circular in cross section.
Terminal. Applied to the end bud beyond which no further growth takes place normally until the following season.
Ternate. Divided into three segments; 3-foliate.
Thorn. A stiff, woody sharp-pointed structure which represents a modified branch.
Tomentose. Densely pubescent with soft, matted hairs.
Truncate. Ending abruptly as if cut off transversely.
Tuber. A thick, short, underground branch or part of a branch, with many buds.
Tuberous. With tubers.
Umbel. An inflorescence with the flowers arising at the same point.
Umbo. A small scar on the apophysis of a cone scale of a conifer.
Unarmed. Without thorns, spines, or prickles.
Undulate. With a wavy margin; repand.
Valvate. Said of buds in which the scales merely meet at the edges without overlapping.

Valve. One of the parts into which a capsule splits.

Verticillate. With three or more leaves at a node; whorled.

Villous. With long, soft hairs, not matted together.

Viscid. Sticky.

Whorl. A group of three or more similar leaves arising from a node.

Winged. (Of stems, petioles, or rachises) Flattened structures projecting out along the sides.

INDEX

Names underscored are synonyms. Numerals underscored indicate pages on which illustrations appear.

C

I

J

K

L

M

Q

X

Y

Z